Roy U. Schenk, Ph.D.

THOUGHTS OF DR. SCHENK ON SEX AND GENDER
A Bioenergetics Press Book
October 1991

For information: Bioenergetics Press, P.O. Box 9141, Madison, WI 53715

ISBN 0-9613177-2-8

Printed and published in the United States of America

Thoughts of Dr. Schenk on Sex and Gender

Roy U. Schenk, Ph.D.

Preface

This book contains ideas developed and written during more than 23 years. A major part of the book involves quotations from the original sources; the book is made up of a series of quotations on each topic. Occasionally an update is added parenthetically. It is not written to please women, so it uses words not considered appropriate around "the ladies". I make no apologies for this. I refuse to treat women as superior and so as deserving of special consideration.

During the seventies, I became vividly aware of a guilt response by men which appeared to be a response to being male rather than to any specific inappropriate behavior. I called this Male Existential Guilt. A workshop participant remarked to me that this was the definition of shame; and I later confirmed this, i.e. guilt is related to behavior (I did something wrong or I failed to do something right), whereas shame has to do with one's self (there's something wrong with me). So I renamed this response the Shame of Maleness—a sense of shame for being male. In this book I have changed the word "guilt" to "shame" in the quotations from earlier writings where appropriate in order to identify the correct effect.

It is my hope that the book will generate far more light than heat, though I do recognize that any man who strives to present a male perspective will be subjected to intense attacks. I recognize that when a group is seen as inferior, that any member of that group who says or writes anything critical about the behavior or attitudes of the "superior" group will be attacked and condemned by the presumed superior group and their lackeys. And it is quite evident that men are seen as inferior to women in relationships, morality and other ways. So, as I have learned over the past two and a half decades, a man who says anything critical of women's attitudes or behavior will get a lot of heat regardless of how enlightening his ideas may be. There's little one can do about that, so we'll just have to live with it for now.

Several ideas introduced in the first chapter are discussed in more detail in later chapters, so you may want to turn to the appropriate chapter right away.

A number of quotations come from articles reprinted in Dr. Schenk's two earlier books, **The Other Side of the Coin** and **We've Been Had** (see information in back of book). When this happens, it is noted by the letters **OSOTC** and **WBH** respectively in the citation source.

Dr. Schenk's Thoughts on:

Chapter 1: It Can't Go On Like This

Why loneliness is increasing for most of us.

Most of us suffer from intense loneliness even while we yearn for a close relationship with another person. Most of us would like to have a close and caring relationship with a person of the opposite sex. Yet there seems to be so much antagonism between men and women that the probabilities of such a long term relationship developing seem very bleak.

Unfortunately, one of the major causes of this intensified antagonism over the past couple of decades involves a sacred cow most people dare not confront. The source of this antagonism is the feminist movement's use of blaming and shaming men to attain their objectives. Since men's negative behavior towards women is induced primarily by men's feelings of inferiority towards women and this blaming and shaming intensifies those feeling of inferiority, the feminist movement has actually aggravated the problem rather than helped to heal it.

In addition, feminists have been using shame to get governments at all levels to pass draconian anti-male legislation which can only cause increased mistrust of women by men. I find it difficult to believe that this is what most women really want because it is increasing the alienation between men and women and so is increasing our loneliness and isolation.

Writings, 8/91

Can sexism be cured with sexism?

We've been hearing a lot about how men need to be sensitive to women's feelings and women's needs. But when was the last time you heard that women need to be sensitive to men's feelings and men's needs? Indeed, other than anger, when have you heard anything at all about men's feelings and needs?

What we have been hearing for too long is about women's rights and about men's responsibilities. Where is the justice in that? Where is the fairness? This is sexism pure and simple. When are we going to hear about men's rights and women's responsibilities? What we are experiencing is the sexist attitude that women have rights and men have responsibilities which often creates the view of women as victims and men as perpetrators. What we encounter is the sexist attitude that men don't have feelings or, if they do, that those feelings don't count.

What we encounter is the sexist attitude that women have a right to be treated with special privilege and that they should not be held responsible for how their attitudes and behavior affect men. What we encounter is the sexist attitude that violence to men is unimportant and indeed that men deserve any violence they experience. (See chapter on violence.)

What we have is an intense level of sexism—posing as anti-sexism. This sexism dominates our college campuses, it dominates government agencies, it dominates newspapers and the media, it dominates our work and our play. In short we have a society that is saturated with this sexism while promoting the lie that it is opposed to sexism. This is a sexism that puts men down and aggrandizes women, that sees women as inherently superior to men, that is supported by most feminists even though it is a total contradiction to feminists' claims that they are merely seeking equality with men.

Writings, 9/91

J'accuse.

I accuse the leadership of the feminist movement of:

1. Being more against men than for women.
2. Having little or no interest in equality between men and women.
3. Cynically speaking about equality primarily to foster their own power.
4. Intensifying the conditions which cause rape.
5. Wanting more women to be raped so they can get more anti-male legislation passed.
6. Having little or no concern for the long term interests of most women, particularly

women's desires to have close relationships with men.

Writings, 10/91

New insights needed.

Certainly the situation as it exists now cannot be termed satisfactory. We see a great deal of lip service by men in positions of power to equal opportunity, to equal pay, to equal political input. Yet women's average pay relative to men's has not increased over the past decade—if anything, it has declined. Supposed egalitarian relationships between men and women are too often not working, with women as well as men being unhappy with them.

What's wrong? Why are things going so badly? Can we still continue to maintain that these problems are all men's fault? What fundamental problems are there that we may be overlooking? Are there new insights which can help us to make more sense out of what is happening and help us to make the struggle easier and/or more successful?

I believe what I am saying here can help provide answers to some of these questions.

The Other Side of the Coin, pp. 14–15, 1982

Men and women both oppressed.

Promoting women as the sole, or primary, victims has also permitted women to attain and maintain a privileged status which puts them above criticism, especially by mere males. In reality it is becoming evident that men are at least equally oppressed and victimized, though obviously in different ways than women are oppressed and victimized.

Statement to Wisc. Equal Justice Task Force, 12/89

Male perspective needs to be heard.

Feminism has raised our consciousness, but this has been one-sided consciousness-raising. If you take a plant, and put it in a window where it gets sun only from one direction, the plant bends and distorts towards that light. You need light from both sides to keep the plant growing healthily. In a similar way, what we need in order to have healthy male-female relationships is to have a male perspective as well as a female perspective articulated and understood.

Speech, 1/91

You can be you and still get laid.

I've observed over the years that there are a lot of men who are afraid to confront women on gender issues. Indeed there will likely be many men who will be afraid to read this book or to talk about the ideas and insights they learn here. They are afraid they will never again get any sex from a woman if they do. Does that tell you something about the immensity of women's power? My experience has been that some women will indeed reject you, but there will be other wonderful women who will be delighted to find men who are finally willing to relate to women as equals rather than as men who are downtrodden inferiors.

One hint: Don't judge women primarily by traditional beauty standards. Women who depend on beauty to get by are less likely to be empathetic. Look mostly for their interior beauty, and don't seek partners among women who focus primarily on women's rights.

Writings, 9/91

Those privileged are protected.

We need to recognize that protection is extended to important persons, e.g. the president and other high government officials, rather than to unimportant persons. Men also extend protection to more important people, i.e. women and children. "Women and children first!" on lifeboats, etc., says that women and children are more important than men. It's never "Blacks and Chicanos first!"

Time for a New Phase in the Liberation Struggle, 1975, OSOTC, p. 221

Men's perspectives resisted by feminists.

I have gotten some positive feedback to my ideas from a number of men. I have also gotten some near violent reactions from some women. This latter reaction I have come to expect for anything I write which deviates from the standard feminist perspectives. I guess we men are to be liberated on women's terms, or not at all.

Time for a New Phase in the Liberation Struggle, 1975, OSOTC, p. 221

Men's inferiority overlooked.

Feminists have been saying for decades that women are seen as inferior to men and that this does devastating damage to women. What they are saying has some truth in it, but this inferiority is limited to the material areas such as the economic and political.

What has gone unsaid for too long is that men have been perceived by society as inferior to women in what I call the spiritual areas, which include morality and social graces as examples. I've got news for you, fellows. As long as women are seen as superior to men in these areas, we will be fighting a losing battle.

Men Morally and Spiritually Inferior to Women?, Legal Beagle, 8/86, WBH

Men shaming men.

The men in Men Against Rape groups are filled with the **Shame of Maleness**, and their shame drives them. So when they talk to other men, or to boys in schools, they have an intense need to pass on this shame to these other men and boys. This is their way of trying to get rid of that shame. This shaming by these men is counterproductive. It actually contributes to the more frequent occurrence of rape because it is highly shamed men who tend to commit rape.

In order to reduce rape, men must be

treated like human beings and not be made to feel shame about their maleness. It requires helping men learn to love themselves and other men since shame results from feeling unlovable. There are ways to facilitate learning this love. Almost certainly this requires belonging to a group of self-affirming men who consciously strive to avoid putting down and shaming themselves, each other and other men and who strive to help each other heal their childhood wounds, particularly their shame wounds.

Writings, 9/91

Men's wounds.

What are men's wounds? First, the message that because we are male, we are bad. Another wound is from the emotional abuse that men experience throughout life, including put-downs relating to clothing, to language, to vulgarity, to eating habits, to sexuality, to our being more physical than women are.

Women's behavior is set up as the standard for good behavior, and we men get the message that because we are different, we are bad. We learn to feel ashamed of being male, including shame about our sexuality. This produces machismo behavior, and it gives women tremendous power to control and manipulate men. Women's controlling and manipulating of men is the ultimate wounding experience.

Speech, 1/91

Men must choose changes that benefit themselves.

Men have a great deal to gain from an egalitarian, non-chauvinistic society—such things as a longer life-span, reduced usage as cannon fodder in wars, a lower percentage of males incarcerated in jails and prisons (about 95% of prisoners are now male), greater availability of sex, and improved self-image that will remove us from the perpetual role of the heavy and being viewed as the violence-doers. It will liberate us to delight in our sensuality, our gentleness and tenderness, our needs for dependency, our feelings in general, and our other so-called "feminine" attributes presently denied us. Feminists aren't going to do this for us. In fact, their continuing use of guilt and shame helps to perpetuate our bondage. We men are going to have to rise above these feelings of shame and guilt at being men and develop our own perceptions of how we are to liberate ourselves.

The Other Side of the Coin, pp. 15–16

Historically women have rejected equality for power.

Like other historians, feminist historians maintain that women have dragged men out of barbarism into civilization. It certainly is clear that women have claimed for themselves the role of civilizing men and of maintaining the finer things of life. Feminist historians also insist that prior to civilization there was an egalitarian society of "goddess worshipping men and women". Why then did women drag men out of this egalitarian society to a society where, feminists insist, women are subservient to men? The answer is surprisingly simple. They didn't. They dragged men out of this egalitarian society into a society where women have power without responsibility, where men do women's violence for them, where women control and dominate men while shaming men for their violence and "male dominance", and where women teach men that their proper role in life is to maintain and increase the standard of living for delicate, superior women.

Letter to The World (unpublished), Oct., 1988

Language reflects social expectations.

Men react in an intensely negative way to such accusations as "You're a sissy!" or "You're just like a girl!" Women on the other hand react equally as negatively to such accusations as "Slut!" or "You're a whore!" In each case they are reacting to a message that says they do not measure up to society's standards for their gender. Another way that this is manifested is in the use of the word "tramp". A tramp is a person who does not measure up to society's role expectations. A male tramp is a person who doesn't work as much as society says he should to support his special woman and children. A female tramp is someone who has sex more freely than society says she should and so reduces its market value for other women.

The Other Side of the Coin Revision, 1991

Do male primates show Shame of Maleness?

The attitudes of superiority that men and women learn undoubtedly had to develop together and developed far back in antiquity—in fact probably during the evolution of humans from primates. Actually, I sometimes think I can see shame expressions in the face of the male orangutan at the zoo.

Moral Superiority—The Real Psychic Source of Misogyny—Letter to Christianity & Crisis, 1979 (unpulished), OSOTC, p. 235

Male beasts don't deserve good bathrooms.

Women as angels, men as beasts, women as morally superior, men as degraded and morally inferior is the message that permeates our culture. Just as blacks and other minorities (also women in material ways) have been damaged by the social attitude that they are inferior, so are men damaged by this attitude of their imputed moral (also cultural, social, spiritual) inferiority. Thus, for example, if one compares public restrooms for men and women, one finds that women's facilities are often far finer.

In the southern U.S., in the past, the finer facilities were provided for white people; and blacks and other minorities got the less fancy, even crude facilities. The reason for this was that blacks were perceived as inferior to whites. Men are given the less fancy facilities for the same reason. By itself this would not be important, but it is not an isolated instance, it is simply a part of a consistent pattern that sees men as inferior.

Guest Editorial, Capital Times, 12/10/87 and The Other Side of the Coin, p. 30

Inferiority is original sin.

The belief in "original sin" prevalent globally is the belief that men are inferior to women in relationships and women are inferior to men in achievement. These beliefs are based on societal value judgements since there are no absolute superiorities and inferiorities. These society based inferiority/superiority beliefs are lies which wound us all, not only in our relationships with people of the other gender, but also in our own personal self-image. In short, a critical part of our world view is based on a lie which damages us all.

Writings, 3/91

Are there acceptable stereotypes?

In its final report, the Wisconsin Equal Justice Task Force discussed in detail and at great length the serious negative effects of stereotyping. The task force then produced a report which accepted and supported four of the most dominant stereotypes prevalent in our society.

These four stereotypes are: First, women are victims and men are perpetrators. Second, women should be treated with special privilege, and women should not be held responsible for how their attitudes and behavior affect men. Third, violence to men is unimportant, and indeed men deserve the violence they experieñce. And fourth, men don't have feelings, or if they do, their feelings don't count.

Report on Wisconsin's Equal Justice Task Force, 9/91

Men screwed by trying to screw.

Unless you buy into the sexist belief that men are inherently more evil than women, it must be apparent that there is something severely wrong when so many well respected men end up being exposed and disgraced for seeking to meet a very human need (sex) in ways society has defined as inappropriate or illegal.

What is obvious to me is that women are behaving in equally inappropriate ways but we do not make these behaviors immoral or illegal. The reason for this is that women define morality, and laws follow morality. What we make illegal are the behaviors men tend to pursue, such as prostitution, which is largely a result of the sexual deficit which men experience throughout their lives. We do not punish the flaunting, enticing, provocative, abusive behaviors that women do as a result of their sexual surplus.

Statement to Wisconsin Equal Justice Task Force, 12/89

Is there hope?

Today, relationships between men and women are often antagonistic, and a great many people believe the situation is getting worse. So—is there any hope of reversing the antagonism and attaining the trustful and delightful relationships between men and women that most of us desire and dream about? I believe the answer is yes. But first we need to understand the causes of the problem. Modern medicine has succeeded in curing many diseases once it has learned what causes the disease. And we can have hope that we also can cure the disease creating animosity between men and women once we learn the causes.

This book identifies these causes and offers some limited ideas on what we can do to remove them. But a great deal of dialogue between men and women is needed to fully develop the means to correct the problem. An essential first step requires that women accept men as equals with equally valid ideas, that women recognize that they are equal contributors with men to the problem and to the solution as it develops and that both men and women give up attitudes of blaming. Instead of blaming we need to ask each other to accept responsibility for our respective contributions to the problem; and we need to honor and celebrate the changes we make as we mutually resolve the animosity and move to trustful and caring relationships.

Writings, 10/91

Chapter 2: Equality

Men's views equally valid.

Please note that I accept women's liberation viewpoints as valid viewpoints—equally as valid as men's liberation viewpoints as these emerge, but not superior to them.

Time for a New Phase in the Liberation Struggle, 1975, OSOTC, p. 221

Judgements of inferiority cause oppression.

It is important to realize that differences between people, whether they be gender based or merely individual inherent capability differences, are not intrinsically a source of discrimination. It is only when a judgement is made that the differences make one person or group superior to another person or group that discrimination occurs.

It is the belief in the inferiority of a person or group that permits one to treat that person or group in a derogatory, condescending or contemptuous way. It is my contention that men are treated in derogatory ways because we are seen as inferior to women in the relationship or "being" areas of living, including emotions, caring, nurturing, and of course morality.

Statement to Wisconsin Equal Justice Task Force, 12/89

Equality of long term relationships.

A psychologist whom I consulted in the past said on several occasions: "In a long term relationship, I'm convinced that things come out about equal." I agree with this idea—that in a long term relationship, each person encounters about an equal amount of met and unmet needs. The proportion of the two may vary from relationship to relationship.

I often allude to past equality between men and women as an equality of dumping—women dump on men and men dump on women an equal amount.... We have no valid justification to spend our time guilt-tripping. Instead, we should examine human liberation from a man's perspective.

Human Liberation: Facing Up to Our Responsibilities, 1976, OSOTC, p. 232

Feminists confuse self-interest with equality.

It is futile for feminists to demand of men that we give up our areas of superiority while they continue to maintain their "spiritual" superiority. It is essential that feminists get off their pedestals and abandon their other superior attitudes and behavior. One way of doing this might be for them to quit promoting themselves as working for "equality" when what they really are doing is only pursuing their own short term best interests, such as seeking better paying jobs and more political clout.

If feminists are serious about wanting to

join men in working for equality they will have to take a broader, more long term view of their own best interests. I operate under the assumption that both men and women are working on their own long term best interests when we all are working for real equality, not just the removal of our disadvantages.

The Other Side of the Coin, pp. 186–7

Feminists destroyed equality.

In the old intersexual dynamic, there was a real tradeoff and equilibrium between men and women which produced a kind of equality between the sexes. Feminists' demands for greater political and economic power have disrupted this equilibrium and little attention has been paid to the compensating benefits men can reasonably expect to receive as women move towards economic and political equality. Feminists have told men what their supposed benefits would be, but there has been little tolerance for differing male viewpoints, especially those which suggest that women must give up some advantages also.

The Other Side of the Coin, p. 19

Do men deserve equality?

What is amazing is the intensity of the negative reactions one encounters to the suggestion that men deserve equal treatment and equality with women. These negative assaults are not restricted to militant feminists, but a great many of them certainly lead the pack in attacking such a "preposterous suggestion."

Writings, 9/91

Same game, different rules.

Actually, it is only the positive male gender terms that feminists insist on degenderizing. Negative terms, such as "patriarchy", are utilized with accelerating frequency. Mary Daly, In **Gyn/ecology**, even used numerous newly-coined words such as "phallocracy", androcratic", and "phallotechnology" to put men down. And positive female gender terms are still held on to. For example, the **League of Women Voters** solicits membership from both men and women but still retains the original sexist name. In effect, feminists are demanding half of the glory but refuse to accept half of the responsibility or blame.

The Other Side of the Coin, p. 7

Does "equality" leave women one up?

Providing women with equal job opportunities, etc., does not produce equality. It only puts women in a one-up position. I hear feminists arguing that it's about time women got one-up for a change. But in reality, given the forms of inequality experienced by men and women, I rather believe women have always, overall, had a somewhat one-up position. And I believe this is an important part of why there is such intense resistance to the present efforts to provide equal economic and political opportunities.

The Other Side of the Coin, p. 20

Will women lower themselves to equality with men?

I personally believe that men's areas of perceived superiority are poor solace for men's perceived moral/cultural inferiority. And if this is so, if the present status is bad for both, then it's time for a change—a change in the attitudes of both sexes, not just in one or the other.

Human liberation and equality, I am convinced, cannot happen until men's feelings of shame at being male are eliminated. But this can only occur by women giving up their perceived spiritual superiority, by "lowering themselves to equality with men."

The Other Side of the Coin, p. 68

Feminists gain power by claiming to promote equality.

Feminists are demanding that we men give up our areas of perceived superiority and special privilege; but they have failed to address, indeed regularly fight to retain, women's perceived spiritual superiority.

If actions speak louder than words, and if one judges feminists by their actions, then a lot of feminists are far more interested in power than they are in equality. They do give lip service to equality and indeed far more than this—they are vociferous advocates of equality. But this does not detract from the reality that their actions show them first and foremost to be power-hungry. Unfortunately, it does greatly muddy and confuse things.

The difficulty is that a great deal of their power is dependent on their seeming to be advocates of equality. This retains their "spiritually" superior status, i.e. the idea that they aren't really in it just to increase their power. And this gives them a great deal of power. If they acknowledged that they aren't really above such things, that they do indeed lust for power, they would lose the substantial power produced by their projected spiritual superiority.

It is expecting a great deal to expect power-hungry persons to willingly give up their power base. But I do feel it's unfortunate that our struggles for equality are impeded by these vociferous feminist "advocates." If it were not such an important struggle we could tolerate it more willingly. But since I have the notion that this is a life-or-death matter, I continue to be very upset by feminist abusers of "equality-advocacy" to gain greater power.

The Other Side of the Coin, p. 119

Feminists want benefits but not responsibilities.

It is not clear to me that many feminists are really seriously interested in equality. This may sound harsh, but it is a conclusion based on over a decade of observation and experience. There is no doubt that feminists are vehemently interested in eliminating their own and other women's disadvantages, such as low paying jobs, impeded educational opportunities, and inadequate political input. Actually, non-feminists are also usually interested in getting better pay if they must work.

On the other hand, I have seen little evidence that feminists are willing to share the harsh realities men experience. I see little evidence of a willingness to give up their perceived spiritual superiority and all that goes with it: such things as being protected by men; avoiding the threats of military combat; being able to blame men and to use that blame to manipulate men; and retaining the advantages of their sexual capital and the power they achieve through making sex unavailable for men and expecting that men will take the initiative in relationships, especially in the areas related to sex. Also an awful lot of women still appear willing to trade away economic liberation in return for the pampered slavery of having a man to support them.

Likewise, I find a willingness on the part of many women, feminist or not, to demand equal pay for jobs and still to expect men to pick up the dirtier or more physically demanding parts of the job. This translates to equal pay for less work. This generates intense resentment in men, even while they are being beautifully manipulated by the women to take on the onerous tasks. These men know what is happening is not equality and they will be increasingly reluctant to support women's efforts in the future.

Although feminists demand an equal share of authority for women in organizations, too often there is the tendency to pick and choose what responsibilities they will accept. And there is a strong expectation that men will pick up whatever responsibilities women choose not to accept. Somehow, that doesn't seem like equality either.

The Other Side of the Coin, p. 130

Are there feminists committed to equality?

One area where women brag about their superiority is in relation to the greater numbers of their sex who are active in what they perceive as the struggle for liberation and equality. I say "perceive" because I am not certain I've yet encountered a woman really committed to equality. For certain I can count their numbers on one finger. Struggle to eliminate women's disadvantages and disabilities? Yes, of course! But willing to give up their special privileges and manipulation of men and to struggle to help men to eliminate their disadvantages and disabilities? If so, it is only on the basis of their own perception or bias rather than trying to see these problems through men's perceptions or bias as well.

Time for a New Phase in the Liberation Struggle, 1975, OSOTC, p. 221

Feminists' oppression of men does not create equality.

Equality for women cannot be achieved by increased oppression of men—by increasing men's perceived moral inferiority and resultant shame feelings. Not only is this a contradiction in terms, but it is psychologically necessary for men to respond, and to compensate by resisting, and by retaining and bolstering their areas of perceived superiority.

There is another way—a way that can achieve equality. It is through the elimination of women's perceived moral superiority. For when men's perceived inferior moral status is removed, they will no longer need to keep women down in other ways.

This will require an almost 180° reversal in the present attitudes and approaches of feminist leaders. It will necessitate building men up instead of continuing to put men down. It also will require feminists to come down off their pedestals of self-righteous moral indignation.

First Take the Log out of Your Own Eye, AHP Newsletter, 4/79, OSOTC p. 238

Is there hope?

In recent years there have been a small but increasing number of women who have been willing to listen and to dialogue with men in a caring way, striving to treat men as equals. This development offers the greatest hope that we can develop better relationships between men and women.

Writings, 9/91

Chapter 3: Doing/Being

Women do to men, who do to women, who do to men, who...

In describing the oppressions that men and women experience, I often speak of the situation as a vicious circle. In the figure below I have expressed that circle as I understand it—based on the recognition that superiority feelings by one gender induce inferiority feelings in the other gender. The other gender then develops compensating superiority feelings in other areas. These in turn induce inferiority feelings in the first gender, and around and around it goes.

One can start at any point and proceed around the circle. If we start at the point of women's "doing" inferiority, this leads to feelings of incompetence and fear which are well documented. Women's feelings of "doing" inferiority also induce them to develop their own areas of superiority which are in the "being" areas. These induce in men feelings of "being" inferiority. This leads to feelings of defectiveness and shame for being male. It also induces men to develop their own areas of superiority which are in the "doing" areas. This induces the feelings of "doing" inferiority in women and so we are back at the point we started from.

The Other Side of the Coin, pp. 66–68

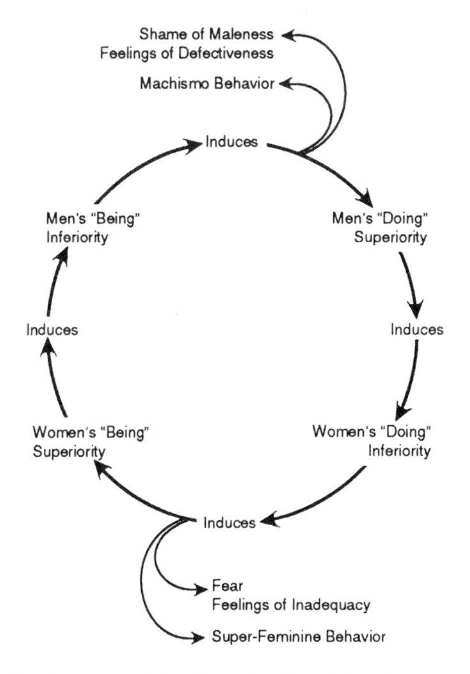

This figure derived from ***The Other Side of the Coin: Causes and Consequences of Men's Oppression.*** *1982. Bioenergetics Press, Box 9141, Madison, WI 53715. $10.00.*

Chapter 4: Socialization

Women teach men they're defective.

What I have become increasingly aware of is that women—mothers, teachers, lovers, wives, etc.—socialize us to perceive ourselves as being defective human beings, as less than human, as animals. This creates feelings of shame; and these feelings are what make men so easily susceptible to manipulation by women.

> *Men Morally and Spiritually Inferior to Women?, Legal Beagle, 8/86, WBH*

Female fear devastates like male shame does.

At the same time that infant boys are being taught to feel defective and so ashamed for being male, infant girls are being taught inadequacy and **fear.** The result is that over 90% of clinical phobias or paralyzing fears are experienced by women. This learned fear is somewhat paralyzing for most women, and it inhibits them from taking the initiative in most life situations including situations in the workplace.

Overcoming this fear is as important for women and as conducive to equality as is overcoming the **Shame of Maleness** for men. This is made extra difficult by the fact that feminist leaders do not want women to overcome their fears, because if they did they would most likely repudiate the man-hating attitudes and policies of the leaders.

Writings, 8/91

Problem starts with women.

Feminists need to accept responsibility for what they and other women have done in their role as socializers. Women have the primary role in socializing; and women not surprisingly plant in the minds of both boys and girls the idea that females are spiritually superior to males. And I believe most mothers also feel it is in the self-interest of themselves and other women to continue implanting in men the idea that men must support and protect women.

I believe feminists have a responsibility to see that these perceptions are no longer transmitted to children by women, and especially not by themselves, and also to reverse the damage done to adult men by earlier generations of women.

The Other Side of the Coin, p. 69

Women reared the men they now blame.

I often think that women's ultimate source of revenge at men for the disadvantages women experience economically and politically is achieved in women's rearing of male children. I consistently scoff at bleeding heart feminists who cry piteously about women's lack of power. What immense power women have through the rearing of children and the generating of shame in men! And how puny men's compensatory economic and political power seems in comparison.

The Other Side of the Coin, p. 141

NOW opposes joint custody to preserve women's control.

Even today, women are invested in continuing to be the primary socializers. For example, the **National Organization for Women** is fighting vigorously to block the increase of joint custody after divorce, even though greater involvement of divorced fathers in child rearing would be in women's best interests, first, because involved men have been repeatedly shown to pay child support and, second, because in joint custody the mother is relieved of the burden of total responsibility for child care. But these reasons don't stop **NOW** from fighting against joint custody, because the desire to maintain women's control of the socialization process is stronger than the desire to benefit women.

Writings, 5/91

Female dominance is lifelong.

Because women are the primary socializers, boys and the men they become develop a **reactor role** to their mothers and then to their female grade school teachers.Ultimately they develop a generalized reactor role to women. This response begins almost from birth and is well developed by the time a child is able to walk and speak. Boys, and the men they grow up to become, learn to **respond** to women and women's attitudes and behaviors.

Because men are trained by women to be responders, women are in a dominant role in male/female interactions.

The result of women's dominance in relationships is that one of the first things men learn, and learn intensely well, is about their inferiority in the "being" or values areas of life. And the message of their moral or spiritual inferiority becomes an overwhelming message for men: a message of shame.

A Revolutionary Proposal for Ending War, 1991

Dampening daughters' desires for sex.

Women do all they can to enhance the differences in sexual need between men and women. And then they use men's greater sexual need to manipulate and control men. Not only do women teach sons to avoid other men (homophobia) and so remain dependent on women to meet their affectional and sexual needs, they also teach daughters to dislike sex. Indeed, in some cultures women cut out their daughters' clitoris to assure a lack of interest in sex or even demand that daughters sexually active before marriage be killed.

The Other Side of the Coin, p. 83

Baby's treatment depends on gender.

Boy babies are touched less than are girl babies from the time they are born. It is no accident that the child's sex is the first question people want answered when a child is born—even before determining if the child is healthy. People don't know how to treat a baby until its sex is known. Studies have shown that a boy baby is handled rougher and less—often left to cry untouched so he can learn to be tough.

The Other Side of the Coin, p. 72

Men's behavior seen as immoral and illegal.

Women have been able to promote themselves as morally superior to men because of their dominant control of the socialization process. Since women have been the primary socializers for millennia, they have naturally promoted their typical behaviors as moral and men's typical behaviors as immoral. This is simply an expression of self-interest, which both men and women tend to pursue, in spite of claims that women are more altruistic. Actually morality tends to be merely the codification of women's self interest.

The shame men feel in response to the socialized message of their moral inferiority has prevented men from directly confronting this anti-equality belief and resulting discriminatory treatment. Since laws follow morality, men also are the losers in the legality game. In other words, the things men tend to do are more frequently made illegal.

Statement to Wisconsin Equal Justice Task Force, 12/89

Is there hope?

It is my belief that most human beings strive for joy and harmony to the best of their ability and understanding. I believe that most women will be willing to give up their domination of socialization and other attitudes of superiority when they understand that this is necessary to develop trustful and caring relationships with men.

Writings, 9/91

Chapter 5: Men's Power

Inferiority driven power.

The power to destroy life on earth many times over now rests in the hands of men—men who are struggling with self-image problems as they lose the superiorities compensating for their "spiritual" inferiority.

Insensitive and unfeeling, these men are carrying the world down the path to nuclear war and the annihilation of life. Intensely burdened with feelings of inferiority, these are men who psychologically cannot afford to lose face in a confrontation. In a nuclear confrontation this could involve being driven to start a nuclear war because their death would be better than dishonor, and if the rest of the world must die along with this supermacho male, well those are the breaks.

The Other Side of the Coin, p. 188

Men losing what little power they have.

There are a few men who have substantial overt economic and political power. Even these men are beholden to women since women own 84% of the country's wealth and women are a majority of voters. So these men know it is only as long as they are "good little boys" that they can continue to run the show. Most men lack this kind of power.

The limited power most men have is the power of earning the family's income—which is then mostly spent by their wives. Men have derived a certain level of power from their knowledge of household and automotive repairs; and finally they often can have a final resort of being physically stronger and so able to physically attack and/or intimidate their wife and children. Such physical attacks are intense inducers of shame in men and so are done mainly by men who are intensely shame addicted. These are men who go through a cycle of shameful behavior, repentance, and good behavior during which time their shame repressed feelings build up, and then repeat the cycle.

Men are losing even most of these areas of power as more wives go to work, as cars become too complex for amateurs to make repairs, and as courts jail men (but rarely women) for domestic abuse. Today professional football and hockey seem to be an outlet for many men to vicariously share a feeling of power since they presently have so few ways to balance women's relationship/moral power over men.

Writings, 9/91

Chapter 6: Female Moral Superiority

Same game, different rules.

I recall my first encounters with the ideas of male chauvinism and women's oppression. I listened. I struggled to understand and to gain a sense of whether these women's complaints were valid. I quickly concluded, yes, if equality means anything, women should indeed have equal opportunities. I also quickly saw that there would indeed be many benefits to me and to men in general from real equality.

At that time I perceived myself as a feminist—a supporter of women's struggles to eliminate their oppression. But that was a short-lived experience. I soon realized that these women blamed us men for their problems and portrayed themselves as innocent victims of us evil men—a phenomenon I now call **The Innocent Victim Syndrome**.

These women had defined chauvinism as an attitude that one group of persons is perceived as superior to another group; and it became quite clear to me very quickly that their attitudes portrayed women as superior to men—morally superior. This is female chauvinism.

The Other Side of the Coin, p. 6

Women's behavior superior.

I suspect that many feminists might well reply: "But of course we women are 'spiritually' superior!" And women do indeed behave that way, just as men also behave as superior to women in leadership abilities, job skills, etc. It's time we recognize that both are equally the result of our socialization, and both must be changed if equality is to be achieved.

The Other Side of the Coin, p. 32

Does women's pain hurt more?

Feminist writer Roberta Lynch, in writing about women's perceived moral superiority, declared that this superiority was poor solace for women's areas of perceived inferiority. (**Press Connection**, 2/2/79, Madison, WI) But then she hasn't experienced moral inferiority, so she cannot know that men's forms of superiority are poor solace to men for our being perceived as morally inferior.

The Other Side of the Coin, p. 35

Betty Friedan nails it.

After speaking and writing on this subject for over ten years to largely uncomprehending audiences, it is with real delight that I recently read in Betty Friedan's book **It Changed My Life**: "If I were a man, I would strenuously object to the assumption that women have any moral or spiritual superiority as a class. This is...female sexism. It is in fact, female chauvinism..."

Yes, it certainly is female chauvinism, and it permeates our culture as deeply and as devastatingly as male chauvinism ever did. Ms. Friedan pointed out that "female chauvinism perverted—and aborted—the first wave of women's revolution in America." She went on to say she hoped this would not interfere with the struggle in the present situation. I'm sorry, Betty, but it already is interfering, and it is interfering very severely!

The Other Side of the Coin, p. 34

Now that we've heard women's side...

The harm to men from men's required leadership and job skills superiority are real enough. Men are expected to support their families. And failure to do so has immense psychological effects on a man. I experienced this intensely myself when I was fired from a job. Men are expected to defend women, including fighting wars, if necessary even to their own death. Men are expected to do the dirty work, to take the risks, to be the pillar of physical strength for those more angelic creatures—women.

On the other hand, the psychological harm to women from men's superior roles in leadership and jobs is certainly real and serious. But feminist literature is replete with descriptions of the lower pay, the self-doubts, low self-esteem, etc. So I am not going to dwell on them. What has not been done and what we do need to do is to take a serious look at how men are psychologically harmed by women's perceived "spiritual" superiority.

The Other Side of the Coin, pp. 43–4

Women seen as the "Good Guys".

Another result of women being seen as morally superior to men is that women are attributed favorable motives, whereas men's motives are assumed to be bad or negative. This is much like the "U.S. =Good Guys, Russia=Bad Guys" attitude of our culture. The result is that if a man and a woman are accused of similar crimes, the man will generally be more likely to be convicted and/or will get a more severe sentence.

Thus women are arrested for 18% of crimes; but only 9.5% of persons prosecuted are women; and only 6% of persons imprisoned are women. This special treatment of women permeates the entire judicial system. The one exception may be "crimes" where the woman threatens women's sexual power by being too freely sexual or by selling her body for too low a price (i.e. prostitution instead of marriage).

Statement to Wisconsin Equal Justice Task Force, 12/89

Valueless women and men.

The attitude that a promiscuous or "fallen" woman is of little value fits quite well into the concept of women's "spiritual" superiority. Since women are these superior beings to be protected by men, a woman who falls from that image is no better than a man spiritually. And since she lacks the economic and political value that men have, she is naturally viewed as nearly valueless, an unperson. The same thing happens to men who choose not to achieve or fail to achieve "material" success. Such a man is viewed as valueless, and most women tend to avoid such men as though they had the plague.

The Other Side of the Coin, p. 41

Chapter 7: Male Shame

Shame based behaviors.

Women's "superiority" areas include morality, and men's sense of moral inferiority induces in men a sense that they are defective ("not good enough") because they are male. This sense of defectiveness translates into a feeling of being unlovable, which is the root source of shame. This shame has a devastating effect on boys and the men they become. One of the manifestations of shame is a striving for overachievement; and boys show this behavior already by the time they are 2½ years old.

Unlike guilt, which has a mobilizing effect, shame has a paralyzing effect. It freezes the person experiencing it. In addition, there appears to be a perception of deserving the bad things that happen due to one's perceived moral inferiority. As a result, men tend to be locked in to their oppression. Once this message of male defectiveness is planted in a boy, the typical male behavior pattern has been implanted, since this behavior is largely a compensatory response to the feelings of inferiority and shame that the boy has learned. Other shame induced behaviors include striving for sexual performance, for power and control, and dissociation from feelings and dependency needs.

Pursuing the Quest for Equality, New Warrior News, 12/89

Men give in because of shame.

How often have you been amazed at the sight of a 250 pound man being "led around by the nose" by a 100 pound woman? How often have you read about male legislators falling all over themselves to pass anti-male legislation because "the ladies" want it? Have you seen men marching in parades or handing out leaflets opposing men's violence to women while totally ignoring women's comparable, more often not physical, violence to men? Have you seen a man give in to a woman without significant negotiations on some important issue, for example child custody? Have you seen a man fall apart when his wife leaves him and abjectly agree to anything to get her back?

If you have seen or experienced any of these behaviors, then you have confronted the **Shame of Maleness**.

Shame and Women's Power, 5/87, Legal Beagle, WBH

Does shame deserve punishment?

If men feel ashamed, then they must feel they deserve to be punished. This then becomes a need to be punished, and their oppression becomes their punishment. We do indeed observe that men often feel they deserve their oppression, and as a result they are unable to struggle effectively to eliminate the oppression. A further result of this appears to be an inability to criticize the "morally superior" women or even to tolerate such criticism by other men.

The Other Side of the Coin, pp. 53, 221

Hard to overcome shame.

Just as women find it difficult to overcome their fear and intense feelings of inadequacy, so also men will find it difficult to overcome that **Shame of Maleness**. It is buried in the psyche so deeply.

Overcoming Men's Fears of Intimacy, workshop at the 16th Midwest Regional Conference of the Association of Humanistic Psychology, 1986

Shame distorts reality.

One of the problems we men have is a lack of models—models of men who are living a life free of Male Shame. Just as I've heard it said that a fish would be the last creature in the world to discover water, so it is that men are so immersed in shame, it seems natural. It is so much a part of our perception of what it means to be male, that it will be very difficult for us to step outside of that reality and see the reality-that-could-be in the absence of our oppressive shame-feelings.

The Other Side of the Coin, p. 175

Double whammy of male shame.

Both men and women feel shame if they deviate from their "gender ideal", e.g. women—beauty, body shape; men—strong, protective of women. Men also feel shame to the extent that they **do** conform to their gender ideal. This shame is a result of the message men get that we are inferior or defective in the values or "being" areas.

This results because women are seen as the standard for these areas, and because they are male, men are automatically assumed to be unable to measure up. This assures that men will feel ashamed solely because they are male. That is what the **Shame of Maleness** is all about—being ashamed of being male. Since it is learned so early, men are not generally consciously aware of these feelings. The shame just seems to be a part of men's lives.

It is this **Shame of Maleness** that permits women to manipulate and control men to get whatever women want from men. Like other women, feminists seem to believe that this manipulative control of men by women is natural rather than a result of men's shame feelings. Actually, this shame men experience and the manipulative control by women are the greatest impediments to equality, to reduced gender violence, and to a peaceful world. As a result removal of the imputed male inferiority and resultant shame has the potential to dramatically reduce male-female antagonism and threats to world peace.

A Revolutionary Proposal for Ending War, 1991

Abandoning blaming and accepting responsibility.

Asking a person to accept responsibility for his or her behavior is both appropriate and caring. On the other hand, blaming and so shaming a person for his or her behavior is dehumanizing and uncaring. Unfortunately, women have learned to use blame and shame as a means to control men to the extent that it is difficult for women to relate to men without sending shaming messages. Also, it is hard for women to believe that a request for responsible behavior is not an attempt to shame them, since women so regularly use blame and shame towards men.

Men are so sensitized to women's shaming that it takes only a small shaming message to produce intense shame feelings. For example, when a woman remarks to a man: "That is a nice shirt you are wearing," this is a shaming message. The message conveyed is: "Because I'm a woman, I am an authority on clothes and as a man you are such a klutz that you wouldn't know, so I'm telling you: that's a nice shirt."

Actually men are so sensitized that they might even get a shaming message if a woman said the more appropriate: "I like your shirt." That may leave women perplexed as to how to talk to men. It may also help women understand that they have an equal responsibility with men for our current problems and that we need to work together in caring ways to resolve these problems. This includes accepting responsibility for how each of us, often unwittingly, contributes to the problems and therefore how we each need to contribute to the solution. We are not powerless. We can make a difference.

One way to make a difference would be to confront the massive blaming and shaming campaign feminists are pursuing about rape. This campaign actually fosters rape because the more shamed a man is the more likely he is to respond violently. What is needed, instead of a campaign of blaming and shaming men, is a program to enhance and build up men's self-image, including a dialogue in which women can come to understand their contribution to the problem, and a mutual acceptance of responsibility to reduce the violence done by both men and women to each other.

Writings, 8/91

Chapter 8: Innocent Victim Syndrome

Victimization and privilege.

A great many women appear to suffer from what I call the **Innocent Victim Syndrome**, the belief that women are innocent victims of evil men. Women seem unable to accept the idea that they contribute to our intersexual problems in ways that have not been consciously recognized (usually these behavior patterns benefit women, e.g. manipulation). And there appears to be intense resistance by women to giving up their special privileges, even while they demand that men give up theirs.

It seems to me that we need to deal, as a fellow in a recent men's group meeting remarked, with the elimination of special privileges of all people and classes. And just as women have demanded that men change on the basis of women's perceptions of the problem areas experienced by women, so also it seems equally valid for men to demand that women change on the basis of men's perceptions of the problem areas experienced by men.

Time for a New Phase in the Liberation Struggle, 1975, OSOTC, p. 231

Is violence justifiable?

In some ways, many feminists are the most blatant sexists, for they often manifest the **Innocent Victim Syndrome** in the most extreme form—the attitude that women are innocent victims of evil men. A couple of dramatic examples of this sexist behavior have occurred here in Wisconsin during the past several years. Two women on differing occasions finally got fed up with their husbands' threats of violence, obtained guns, and deliberately murdered them. These women are held up as heroines by feminists even while these same feminists vociferously deny that a man ever has a justification to respond with physical violence to women's more subtle kinds of violence.

The Other Side of the Coin, p. 32

Males dare not criticize females.

Promoting women as the sole or primary victims has also permitted women to attain and maintain a privileged status which puts them above criticism, especially by mere males. In reality it is becoming evident that men are at least equally oppressed and victimized, though obviously in different ways than women are oppressed and victimized.

Statement to Wisconsin Equal Justice Task Force, 12/89

Are Women's Studies Programs counterproductive?

One of the major goals of Women's Studies Programs appears to be maintaining and enhancing the **Innocent Victim Syndrome** which serves to further blame and shame men. This enhances women's traditional power over men rather than moving us towards equality. It also plays on women's fears and feelings of helplessness and increases women's distrust of men. So these programs may be counterproductive to the goal of encouraging women to become equal partners with men.

Writings, 8/91

Chapter 9: Women's Power

Women must share power they deny.

There was a story in the paper about a year ago about a five-year-old boy who took his dad's pistol and shot and killed a seven-year-old playmate. The five year-old thought the gun was a toy, so he was not aware of what it could do. Yet, in spite of his ignorance, he had an immense amount of power. And, unfortunately, that power was misused in a way devastatingly harmful to the seven-year-old.

Women's power is something like this, I suspect. Women have enormous power over men that was planted there when men are children. Yet many women seem unaware of this power, or at least claim to be unaware of it. And they too often misuse that power in ways that are extremely hurtful to us men.

It is interesting that feminists also usually deny their power or treat it as inconsequential. I see this as a real put-down of women; and I think that the denial of that power, the treating of it as unimportant, is one of the things that prevents feminists from speaking effectively to many women and men.

Perhaps the next essential step in dealing

with equalization of power, and the hurtful abuse of women's power over men, is helping both men and women become aware of this immense power. Beyond that it will be necessary for men to begin demanding that women share this power in exchange for equal economic and political power.

The Other Side of the Coin, p. 121

Surplus sex gives women power.

We need to recognize sexual power. We've all been told how powerless women are economically, as though most of us men have a great deal of economic power. Actually, it is only the wealthy person, male or female, who has a surplus of wealth and therefore economic power. I don't know about you, but that certainly excludes me. Those of us with a lack of wealth have a deficit of economic power. After all, that is what economic power is about. Can we now apply these same concepts of surplus and deficit to sexuality?

Sex will almost never happen the first time between us and a woman unless we initiate it; and we are turned down far more often that our offer is accepted. When we are turned down, we have a sexual deficit, and the woman has a sexual surplus.

Just as those persons with surplus wealth have economic power, so also those with surplus sex have Sexual Power, and **that is women**. This sexual power is far more broad-based than whether you get laid tonight. One of the most important ways women retain their sexual power over men is by socializing men to be almost totally dependent on women for their sexual needs, their affectional needs, and yes even their touching needs. After all, who is it that benefits most from men's homophobia? Obviously it's women.

We've all been told about how we men have the sexual power. We've also been told that we've got the economic power, and we know how untrue that is. To put it bluntly, **we've been had again!**

Men Morally and Spiritually Inferior to Women?, 8/86
Legal Beagle, WBH

Shamed men manipulated by women.

Manipulation of men is a way of life for women. It is learned so early that it is just taken to be natural and unchallengeable. It is learned right along with keeping her skirt pulled down over her legs and being dainty. But such manipulation can only work because men also have implanted in them feelings of shame at being male.

If we are serious about striving for the elimination of gender-derived power, then we must address the sources of this male shame—men's perceived spiritual inferiority to women. I see no other way that men can escape this source of domination by women.

The Other Side of the Coin, pp. 111–12

Women deny their power.

I think one of the reasons women want to deny their power is because power carries with it an expectation of responsibility. And women are so accustomed to putting the responsibility on us men, and blaming us when things don't go properly, that even most feminists continue in that same mind-set.

The Other Side of the Coin, p. 118

Women have the choice, they choose the easy way.

Why should women exert themselves through achievements when they can get by simply by embellishing the assets which they have, namely their appearance and their bodies? And indeed few women do choose to base their sense of self-worth on achievement. That is why there are so few major women painters, artists, architects, engineers, industrial or political leaders, the whole bit.

Of course, as usual, feminists have blamed us men for "holding women down" even though "not choosing to achieve" is the preferred choice of most women. And let's face it, it would be the preferred choice of most of us men too if we had the choice. Feminists, and in fact the whole of society, have been telling us that we men have the power and control. That is simply not true. In short, we have been had again.

What is important to recognize is that we don't have the choice. It is women who have the choice. In other words, women have the power.

Achievement or Existence, 10/86, Legal Beagle, WBH

Men's shame produced by and benefits women.

Since women derive a great deal of power from manipulating men's shame, I conclude that men's feelings of moral inferiority and **Shame of Maleness** are implanted by mothers, grade school teachers, and by the entire society. But of course both boys and girls have far more contact with women than they do with men when they are young, even though men also have bought in to the female chauvinist attitude of female moral superiority.

Since this shame permits women to manipulate and control men, it gives women immense power over men. But I find that feminists, and indeed most women, tend to deny this power. I suspect that this is because this is one of women's main sources of power, and they do not want it to be recognized and challenged.

Shame and Male Oppression, 12/87, Transitions, Vol. 7, #4, WBH

Feminists seek pseudo-equality.

I say, let women achieve; but let them come down off their pedestals, share the power of socialization, and give up their sexual power. Then we men have some hope of attaining equality with women. We certainly have little to lose from real equality. It is only the pseudo-equality that feminists preach—where men give up their advantages and women retain theirs—that we need to beware of.

Achievement or Existence, 10/86, Legal Beagle, WBH

What can we do?

It is important for us, both men and women, to develop our own power. This means developing our ability and confidence in controlling our own lives to the extent that is possible. What is equally important is for us to learn to give up our power over others. For men this includes giving up economic and physical power over women. For women it includes giving up sexual, moral and relationship power over men. In this way we can develop caring and delightful relationships between men and women.

Writings, 9/91

Chapter 10: Feminism

Supporting feminism.

I support equal rights, equal opportunities, and equal responsibilities for everyone, regardless of race or gender. To the extent that feminism supports this, I support feminism.

Speech, 1/91

Feminism has become anti-male.

It is sad to see how feminism has degenerated into being more of an anti-male crusade than a women-supporting effort. For example, NOW opposes joint custody for children of divorce even though studies regularly show that fathers consistently pay support for their children when they are involved with their children, as happens in joint custody. Also with joint custody the mother is relieved of the overwhelming 24-hour-a-day, 7-day-a-week, 52-week-a-year total responsibility for the children.

Another example of feminists being anti-male is feminists' efforts to deny a man the ability to defend himself in rape trials by preventing evidence of the woman's behavior and dress. This is also demeaning to women because it treats them as children who should not be held responsible for their own behavior while denying men the right to a fair trial.

Writings, 7/91

Stopping the vicious cycle.

Feminists have been demanding the end of one half of this vicious cycle—women's inferiority/men's superiority. But they have done little to address the other half—men's inferiority/women's superiority. Actually they blame the entire cycle on us men. Also they have accentuated men's sense of spiritual inferiority by playing on men's feelings of shame to induce men to give them what they want.

So it is up to us men to help other men become aware of and surmount the **Shame of Maleness** so we can cease being seen as second class citizens just because we're male.

The Vicious Gender Cycle, 7/86, Legal Beagle, WBH

Men's views needed.

I realize that if I were a woman I would be a feminist, indeed a quite militant feminist—because women do indeed experience some serious limitations and inequities. There are plenty of women addressing women's problems but precious few men looking at men's problems, particularly the way women's socially conditioned behavior harms us men. So it makes sense for me to address these men's issues.

The Other Side of the Coin, p. 5

Denying reality.

I find it very fascinating how feminists so often deny the very things they have complained bitterly about in the past, when it is shown that men experience disadvantages from these same experiences. Thus an early feminist writer demanded to know why women own 80% of the country's wealth [the most recent report is that women own 84% of the country's wealth] and yet men control the economic system. More recently, as the reality became obvious that the person who owns the wealth really does have the power, feminists now deny that women own most of the wealth.

Quite frankly, I don't know whether it is men or women who dominate that small group of people who own the wealth of this country and of the world. I figure most of us don't benefit regardless of who owns it. But we have here another amusing about-face by feminists when they find their claimed oppression can be shown to be at least equally oppressive to men.

The Other Side of the Coin, pp. 108–9

Refusing to accept responsibility.

I find it both amusing and infuriating that feminists have tended to blame us men both for male and female chauvinism, as though women have no input, no responsibility, and no power. It sure must be nice to shrug off any responsibilities for one's own and others' behavior. And so feminists continue to manifest one of women's primary failings—a refusal to accept responsibility for their part in the perpetuation of current social mores, which harm both men and women.

A lot of this refusal to accept responsibility and the tendency to blame men is a natural outcome of women's "spiritual superiority" and the resulting use of shame and blame to control and manipulate men. And it has been quite effective with a lot of men who, oblivious to their own oppression, valiantly and chivalrously devote themselves to the futile effort to eliminate one-half of a vicious circle.

The Other Side of the Coin, p. 124

Socialist feminist dogma.

Many feminists and "male feminists" have come to the conclusion that a combination of feminism and socialism is not only the absolute truth but also the total truth. And they have been promoting this dogma with a religious fervor.

Fortunately this absoluteness does not sit well with at least a modest minority of the social change community in this country. As a result, this new dogma has not totally taken over and perverted the entire social change struggle.

But a major fraction has bought the dogma. For example, the local "alternative" newspaper is openly "socialist feminist", and it will not publish anything I submit to it, or any other writing that disagrees with or questions their dogma. It's almost like the Catholic Church has been cloned.

The Other Side of the Coin, p. 181

Feminists exploit male shame.

One of the basic errors of the feminist movement has been that it takes advantage of the **Shame of Maleness**. Feminists have tried to shame men into doing what the feminists want them to do. If feminists are serious about equality, this behavior is 180° out of sync. They are going in just exactly the wrong direction as far as really working for equality when they use shame to achieve their ends. That kind of behavior is power-oriented, not equality-oriented. If we hope to develop caring relationships between men and women, women are going to have to give up their shaming of men.

Overcoming Men's Fears of Intimacy, workshop at the 16th Midwest Regional Conference of the Association of Humanistic Psychology, 1986

Chapter 11: Morality

Morality is what women do.

Women define as moral what women are inclined to do and as immoral what men are inclined to do. This is illustrated dramatically by the condemnation of pornography, which men enjoy, while no like condemnation is attached to the female equivalent—romance novels and TV "soaps."

What is Men's Wound?, M/R Magazine, 11/86, WBH

Female morality dominates.

We need to challenge the ingrained belief that women's morality is superior to men's morality and that women are inherently more moral than men. The reality is that society's morality tends to be a codification of women's self interest. Society, socialized by women, accepts as appropriate the behaviors women tend to do and as unacceptable the behaviors men tend to do. An example of the codification of women's self-interest as "morality" is the aggrandizement of marriage and weddings, a distinctly female priority, and the condemnation of sexual variety, a more often male priority.

We men also need to work on developing

and honoring our own masculine morality. Such ideas as: "My word is my bond" and "Don't hit a man when he's down" are examples of men's morality. We need to do far more to expand and understand and honor men's morality and spirituality. We also need to work on recognizing that women's morality, culture, and spirituality are not inherently better than men's.

A Revolutionary Proposal for Ending War, 1991

Chapter 12: Violence

Sexist bigotry in legislation.

In our society we ignore the incredible violence men experience—1.6 times as many assaults as women, 2 times as often murdered, and 14 times as often killed on the job according to Federal Government statistics. The list is long. Instead we see Congress debating a Violence Against Women Act to more intensely punish violence against women, while ignoring the 1.7 times more frequent violence against men. If Congress were to propose a Violence Against White People Act to more intensely punish violence against whites, even though minorities experience 1.4 times as much violence as whites, this would immediately be recognized as Racist Bigotry. Yet this Sexist Bigotry bill has hardly raised a protest by those who would vehemently oppose racism, or sexism against women.

MUM Newsletter (Madison, WI), 2/91

Men's rage kept in the closet.

It is time we recognize the reality of men's rage. Let's bring it out of the closet! And let's recognize that it is just as legitimate as is women's rage at men and that its causes are no more shameful or hard to understand than are the causes for women's rage at men.

Women's rage at men is the result of treatment by men which is cruel and violent. And rage is a reasonable response to such cruel and violent treatment by men. Yet, I find that somehow both men and women cannot or will not accept the fact that men's rage at women is also a quite reasonable response to the cruel and violent treatment that we men experience from women.

The Other Side of the Coin, p. 99

Women who dangle sex do violence to men.

Women are socially conditioned in practice to use sex and the covert promise of sex (flirting, teasing, and so on) to control and manipulate men. Unfortunately this manipulative behavior is accepted as normal behavior in our society. But in my opinion, if this manipulative kind of treatment of men by women were directed against any other creature, it would be labeled for what it is—namely, cruel and sadistic. For example, a person who dangled a beef-steak in front of a hungry dog time after time and then yanked it away when the dog reached out for it, would certainly be called cruel and sadistic. His actions would be considered violent, and the dog would ultimately become enraged. But dangling sex in front of sex-hungry men and yanking it away is practiced routinely by women. And rage by men is a most natural response to this cruelty, this violence.

So Why do Rapes Occur? The Humanist, 1979, OSOTC, pp. 242–3

Men rebel against women's control.

Men do rebel against domination by women. But since shame keeps men from overt and direct confrontation of their oppressors, men often respond covertly through passive aggression and hostility, and indirectly through physical violence—an often undirected rageful thrashing out at assumedly superior women—a behavior which itself adds further to men's feelings of shame.

Statement to Wisconsin Equal Justice Task Force, 12/89

Same game, different rules.

There is a strong tendency in society to treat physical violence as much more serious than psychological violence. Thus, when two people are exchanging insults, that is not considered serious. But the first person to strike a physical blow is viewed as the aggressor or the wrong-doer. Since men usually are larger than women and have perhaps more intense and certainly more suppressed rage, they will usually be the first to do physical violence, and so are viewed as the aggressors. [Recent studies show men and women initiate physical violence about equally.]

And yet, if the person who strikes the first blow is a woman, and the recipient is a man, there is still a strong tendency to blame the man and to excuse the woman. In fact, women's violence of the most intense kind tends to be treated as defensive and so as justified. For example, in the past several years in Wisconsin, there have been three highly publicized murders of spouses. In all three cases, involving two dead men and one dead woman, the man has been portrayed as the violator and the woman portrayed as an innocent victim.

The Other Side of the Coin, p. 88

Husband battering not fit to discuss.

Physical violence against women has become a thoroughly discussed subject over the past several years. But the man who comes to work with a black eye inflicted by a woman knows well that not only has this form of violence not been seriously explored, but also that it is a subject that a man dare not discuss openly. A man who is physically assaulted by a woman is felt to have deserved the abuse and/or to be less than a man because it occurred. Consequently, men battering by women is one of the best kept secrets in our society.

The Other Side of the Coin, pp. 93–4

Violence to men unopposed.

What is frightening is the inability of men to rise up and oppose this mounting violence against them. Men's shame at being male is so intense and so paralyzing that almost no objections are raised to anti-male laws. In fact it appears that men's shame makes men feel they deserve the violence they experience. Since feminists have been increasing the amount of violence against men, for example through these punitive laws, it would not be surprising if men's violence against women increased. Instead, men appear to be dissociating themselves more from women.

Writings, 8/91

Even with violence women and men are similar.

I think we men are just as human as women are. And so I think it is just reasonable to presume that we get angry, and ultimately enraged, at the ongoing viciousness, cruelty, and violence we experience from women, just as women are enraged by such treatment from men.

It has to be understood that women's forms of violence are often different than are the violence and cruelty that men do to women. Men's violence and cruelty tend to be more open and physical, whereas women's violence tends to be more subtle, more psychological and emotional. And a lot of it derives from women's power over us as children which generates, by its abuse, intense rage and sets us up for future violence.

The Other Side of the Coin, p. 99

Verbal and psychological violence hurts men as well as women.

Most abusive situations I have observed involved a woman heaping verbal and psychological abuse on a man and the man finally striking back physically. If "the verbal abuse hurts worse than the physical" for women, how about for men, who generally experience far more verbal abuse?

It certainly is true that physical abuse is more visible. As a result, the abuses women do are ignored or are treated as defensive and, therefore, excusable. If the therapist is to take "a clear moral position on the unacceptability of violence in the family," shouldn't that include women's verbal and psychological violence?

Family Therapy Networker, 9/86

No shelters for abused males.

The most under-reported form of violence in the U.S. is husband abuse. Studies by Susan Steinmetz indicate that close to half of spouse abuse is experienced by men. And yet, is there even one shelter for abused men?

Family Therapy Networker, 9/86

Violence by men results from oppression.

We have heard almost ad nauseum about how men exert economic power over women and how women often react violently to that. But there appears to be an almost total lack of comprehension by women, perhaps also by men, of why men can be so sexually violent to women. But if one can recognize that men often behave as a sexually oppressed class, these behaviors make much more sense. Again, let me state that I am opposed to violence, and I do not think men's violence to women is any more justified than is women's violence to men.

The Other Side of the Coin, p. 116

Physical violence to men is not acknowledged.

I have thought in the past that men do more physical violence to women, while women's violence to men is more emotional, mental, and psychological. But I am beginning to have doubts about this past conclusion. With women seen as victims, the physical violence they do to men may simply be excused and so be ignored or even become invisible.

Also, men's macho compensation for their "spiritual" inferiority creates an image of men as physically invincible. This makes it psychologically difficult if not impossible for a man, however puny, to acknowledge having experienced physical violence from a woman. For he would be acknowledging "material" inferiority in addition to his "spiritual" inferiority. When this is coupled with the "he deserves it" attitude, men can become trapped and unable to escape from situations where they experience ongoing physical violence.

The Other Side of the Coin, p. 185–6

Women more vocal.

Women are more vocal about the violence they experience. As a result, the violence men do to women regularly receives more attention and publicity than does the equally vicious but often less visible violence that women do to men.

The Other Side of the Coin p. 87

The choice we have.

Since we men stand to benefit greatly by the elimination of both male and female chauvinism through a better self-image, greater availability of sex, prolonged life-span, and the elimination of other special privileges of women, I hope the choice will be to eliminate the violence done by both sexes—though whether this is possible in a highly competitive society such as ours remains to be seen.

So Why do Rapes Occur?, The Humanist,4/79, OSOTC p. 241

Condemn violence by men and women.

The only way out of the vicious circle of violence between men and women requires us to condemn equally the violence done by both men and women and to refuse to tolerate violence by members of either gender.

The Other Side of the Coin, p. 95

Chapter 13: Oppression

Oppression and liberation shared by men and women.

It is hard to compare the economic and political oppression that women experience more of with the emotional, psychological, and moral oppression which men experience more of. However, I suspect we will never advance very far in the human liberation struggle until either we stop comparing or we assume that both men's and women's forms of oppression are equally in need of elimination and so quit comparing them and just assume that men will work on defining and eliminating our own oppression while women will work on theirs. We must listen to each other when we describe how the other gender's socially conditioned attitudes and behaviors hurt ourselves, male or female; and then strive to modify our behavior accordingly.

Most men cannot openly stand up to the moral self-righteousness and the attitude of moral superiority of feminists. Because of their feelings of shame, men often buy into feminist perspectives and seem unable to recognize that there can be differing male perspectives. They often adopt the attitude that women's oppression is so much worse than is men's. This is typical of the traditional male role in which we are taught to discount and deny our own hurt and pain. The machismo response leads to the military arms race. So the only way to end the arms race is to eliminate men's perceived moral inferiority. Instead feminists try to make men feel even more inferior and ashamed, and so the insanity increases.

Shame & Male Oppression. Transitions, 7(4). 1987

A moral dilemma.

Men's basic oppression, I have finally come to realize after years of study and reflection, is a moral oppression. Men's basic oppression is the oppression of being perceived of as morally inferior to women. This perceived moral inferiority manifests itself as shame feelings, as the macho role, and in the psychological need to keep women down in other ways. It is an immensely powerful and deeply felt oppression, and I believe all men suffer from it whether we are aware of it or not. In addition feminists are at least as involved in this oppression as are other women. Indeed feminists seem to strive to achieve their own liberation by increasing men's oppression. This is not working, but more importantly, it cannot work.

First Take the Log out of Your Own Eye, AHP Newsletter, 4/79, OSOTC, pp. 237–8

Men blindsided by shame.

The perceptions of spiritual inferiority of men, and the resultant shame-feelings, need to be recognized as comparable and equivalent to the material oppressions which are the basic oppressions experienced by women. They are men's primary area of oppression and are just as devastating and just as much in need of change. They are the most serious trauma of the male condition. Unfortunately, men are so immersed in this condition that they are normally totally unaware of its existence or believe it is a natural part of being male. And there are few, if any, effective male models to show the possibility of a different reality.

The Other Side of the Coin., p. 48

Does oppression justify oppression?

A woman associate, in responding to an article I had written about women's manipulation of men, stated emphatically that "If women can be viewed as an oppressed class, then they naturally will use manipulation, lying, stealing, revolution and other 'unfair' tactics in order to survive."

But women are oppressed only in limited ways, i.e. mainly economically and politically. In other ways we men are oppressed, e.g. sexually, emotionally, morally. (Actually, most men are oppressed by both the Economic and the Sexual Ruling Classes.) And so it should not be surprising if men also respond to their oppression by manipulation, lying, stealing, and other "unfair" tactics. In response to their sexual oppression men respond in ways that get them regularly portrayed as cads, as sexually unfaithful, as manipulative.

If we continue the illustration, the woman stated that oppressed people also, at times, rise up in rebellion against their oppressors. Sometimes this happens individually and sometimes as parts of a class. For sexually oppressed men, this rebellion among other things involves rape, including gang rape.

Sexual assault is not, however, any more acceptable behavior than is the manipulative, underhanded behavior of women in response to their areas of oppression. But there is the rub! Women often feel their violent responses to their forms of oppression are justified.

The Other Side of the Coin, p. 114–5

Inferiority means oppression.

A person or group that is seen as inferior is by that very fact an oppressed group. In other words, if we perceive someone as not as good as or as inferior to us, we treat them as inferior and so oppress them. This happens to Afro-Americans and to other racial minorities. It happens to women in the achievement or "doing" areas. It also happens to us men in the areas we are seen as inferior or as not as good. These are the areas of morality, spirituality, nurturing, relationships, etc., the "being" areas of our lives. As a result of women's control of morality, what men tend to do is defined as evil while what women tend to do is defined as correct. As a result of women's control of morality, translated into our laws, 20 times as many men as women are in prison.

How About the 20th Century?, The Liberator, 1/90

A measure of sexual discrimination.

If a person is a member of a racial minority, that person is eight times as likely to be imprisoned as if that person is white. That is a measure of the degree of racial discrimination that occurs in our society. If a person is a member of a sexual minority, that person is 20 times as likely to be imprisoned. This too is a measure of the degree of sexual discrimination that occurs in our society. That particular minority happens to be males. Actually, 75% of autistic infants and 90% of problem children in grade schools are male. This indicates that the problem begins in early childhood and increases in intensity as males mature.

Statement to Wisconsin Equal Justice Task Force, 12/89

Chapter 14: Pornography and Prostitution

Marketing assets.

Pornography basically is an idealized portrayal of women's assets which are most desired by men. The female equivalent of pornography is the romance novels. Pornography, by presenting an idealized portrayal of women's major assets, causes the individual women's own personal assets to diminish by comparison. This is threatening to women because it reduces their value in the marketplace.

Pornography—a Threat to Market Value? 3/87, Legal Beagle, WBH

Women control behavior.

Women are the arbiters, the controllers of morality, mores, and etiquette. Knowing this, it is no longer surprising that pornography, the idealized presentation of what men desire in women, is condemned while soap operas and romance novels, the female equivalent of pornography, i.e. the idealized presentation of what women desire in men, is considered to be acceptable. What we have is a stranglehold by women on the behavior-control system. This is a control far more complete than men's control of the economic/political system has ever been. It is a control that women show little or no willingness to give up.

Women Are in Control, 9/87, Legal Beagle, WBH

Men tolerate violence from women.

One aspect of pornography that upsets a lot of feminists is the violent fringe—the portrayal of violence done to women in a sexual framework. In reality, this violent porn constitutes only a very small amount of the total. In a way that is surprising, since women do such immense sexual violence to men, one would think there would be more depictions of men responding with sexual violence.

Women's sexual violence to men is more subtle. It involves creation of the male sexual deficit, combined with the false advertising of dress, flirting and other provocative behavior, and finally the attitude of moral superiority which promotes women as being above such dirty, animal-like behavior. So it is really rather amazing that men's total response to this intense violence is so limited in its violence.

Pornography—a Threat to Market Value? 3/87, Legal Beagle, WBH

Honest prostitutes.

When a woman expects a man to pay for their dates, she prostitutes herself, in that she uses her sexuality to gain something of significant value (food, entertainment, etc.). In that case the only fair thing for her is to be an honest prostitute. So the only honorable thing for the woman to do is to deliver on the implicit prostitution contract. The woman is also implicitly agreeing that he should be compensated more for his work so he can afford to pay to be with her.

Commentary on Dr. Schenk's Dating Contract, from We've Been Had, 1989

Chapter 15: Abortion

Abortion question reveals attitudes on sex.

Society's anti-abortion attitudes seem really more anti-sexual than anything else. It boils down to: If you fuck, you deserve to be punished. Thus a man would not be expected to support a child conceived by donated sperm and artificial insemination but he would if the child were conceived by coitus.

Actually, since Jenner developed the small pox vaccine and so began the development of death control, the human population has been expanding explosively. The result is a drastic need for birth control. Yet most anti-abortionists are also opposed to birth control. Again this seems to be more of an anti-sex attitude than anything else.

Writings, 7/91

Father's rights lost in abortion debate.

I think often abortion opponents are reacting negatively because of abortion supporters' attitude that only the woman has any rights when it comes to abortion, thus denying any rights to the prospective father and also denying the fetus the gradually increasing rights it should gain over time as it develops. In denying a man's rights, pro-abortionists are inconsistent because if the woman should choose to keep the infant, they insist that the man should be compelled to support the child financially for at least 18 years, even if he had no say in whether the baby was born or aborted. In fairness, the man should have an absolute say in the decision, or he should be permitted to give up all rights and responsibilities for the child if he should choose to do so.

Writings, 7/91

Catholic Church's inconsistency.

The Catholic Church's absolutist anti-abortion stance violates its own theology. One of the important concepts of Catholic moral theology is the responsibility to choose the lesser of two evils. This concept teaches that when confronting two evils, one must choose the lesser of the two. When a fetus at birth would face a future of starvation and misery and would increase the misery of any older children, it certainly is not clear that aborting the fetus is the greater evil. Furthermore, if the Catholic Church were consistent in defending the sacredness of life, it would oppose war as vehemently as it opposes abortion.

Writings, 7/91

Balancing alternatives.

It can be reasonably argued that we cannot be absolutely certain when the fetus becomes a human person, and therefore prudence dictates that we take the safer course and reject abortion entirely. However, we must surely balance against this possibility the probabilities of such things as the misery of the unwanted child and the mental torture of the woman bearing an unwanted fetus. Surely prudence would dictate giving substantially greater weight to the more certain evils.

Let's Think about Abortion, Catholic World, April, 1968, OSOTC, p. 194

Nature is profligate.

Every human abortion is a tragedy since it is the termination of a potential human person. It is true however that sometimes the alternative can be an even greater tragedy, depending on the circumstances of conception and on the conditions the child would be born into. Confronted with two tragedies or undesirable choices, it is morally appropriate to choose the lesser of the evils, so it appears that abortion cannot be rejected as always inappropriate. However, the deliberate taking of the life of a potential human person is not to be taken lightly since this expresses a cavalier attitude toward human life. Still, nature itself is profligate. It produces billions of sperm as well as dozens of eggs for each fetus conceived. And then a substantial percentage of fertilized ova never implant or spontaneously abort. Obviously nature does not hold life or the fertilized ovum to be totally sacrosanct. So are we to condemn nature also?

Abortion—is there a Middle Ground? July, 1985

Chapter 16: Sex

Scarcity creates interest.

First, some basic realities about our society: 1) Men are taught to desire sex and seek it. 2) Women are taught not to desire sex and not to seek it.

The result of these two realities is that: 1) Men cannot get as much sex as they desire and are even often forced to buy it (prostitution). 2) Women get all they want, which often is precious little, and often have sex forced upon them (rape).

Because sex is scarce for men, it can be used effectively in advertising. On the other hand, I've often heard that bakery workers frequently get to the point that they don't even eat pastries anymore (sort of like some women regarding sex, isn't it?). One could hardly attract bakery workers, then, using pastries as an advertising come-on. Likewise, you can't attract women using sex.

So—an effective strategy for eliminating sex in advertising would be to see that sex is readily available to both men and women. This would eliminate the basic cause that permits sex to be exploited in advertising.

Now—somebody plan the tactics!

Eliminating Sex in Advertising, Midwest Quaker Retreat, 7/72, OSOTC, p. 215

Male sexual deficit proved by experiment.

Many women argue that sex must be equally available as it takes just as many women as men. I have taken to demonstrating the reality that exists by suggesting an experiment for these women. I suggest to them that we go to a bar or some similar place, and I will ask six women to go to bed with me while they ask six men. My prediction is that I will get all negatives, while they will get at least one acceptance, probably the first man they ask. We've never actually tried the experiment because they've always conceded the point after a few moments of thought.

The Other Side of the Coin, p. 79

Women duck responsibility for sex.

Women's attitudes of moral superiority lead them to deny their sexuality and their interest in sex. The result is that they usually leave totally up to men the initiation of sexual activities, particularly the first encounter. It also means women, particularly teenage girls, often do not protect themselves sexually before a date because to do so would be to admit their interest in that "dirty behavior."

Guest Editorial, Capital Times (Madison, WI), 12/10/87

Men seriously under-touched.

While most people in this aloof society of ours receive inadequate touching, males tend to receive far less touching than females do. As a result, men exist under serious deficit conditions for touching. As they mature, men find that the primary legitimate source of touching is in sexual intercourse. This creates an intense association between touching and sex, so that when a woman touches a man he may immediately associate this with a sexual interest.

The Other Side of the Coin, p. 72

Sexual ruling class.

The concept of women as the sexual ruling class needs further elaboration. Just as the economic ruling class is distinguished by having a surplus of money, so the sexual ruling class has a surplus of sex. By contrast, the sexually oppressed class (men) often have to buy sex from their oppressors (prostitution). And women learn early and well to use overt and covert sexual come-ons to control and manipulate men.

Time for a New Phase in the Liberation Struggle, 1975, OSOTC, p. 227

Men's response to sexual power.

Sex is usually the last thing a woman is willing to share with a man. As a result, sex becomes far more important to the man than is touching and affection. Moreover, feminists regularly complain about men's crudeness, focus on self-gratification, and lack of sensitivity to the female partner's sexual desires. But as long as women use sex as a source of power (e.g. "Did you **give** him any last night?") they are demanding too much to expect men to do anything more than to gratify themselves.

The Other Side of the Coin, p. 83

Women's sexual violence.

So we have women creating conditions where men become totally dependent on women for sex; and then women dangle sex and jerk it away and put men down for having such animal needs. They use sex to manipulate and dominate men. In so doing, they do an intensity of violence to men far greater than they evidently realize.

The Other Side of the Coin p. 103

Women in control.

It is interesting to listen to our own language about sex. For example, I have heard women talking among themselves in the morning, and a not infrequent question is, "Well, did you give him any last night?" And among the men I hear, "Well, did you get any last night?" Obviously the person in a position to give or deny is in a power position relative to the person trying to get, as teenagers well recognize when they ask for money from their parents.

The Other Side of the Coin, p. 113

Taboos developed so women can control men.

Anthropologists report that all societies have some form of restrictions or taboos relative to sexual behavior. According to these anthropologists, there is no consistency in what the taboos or restrictions are. But there is a consistent **purpose** for these restrictions. They all create a deficit of sex for men. This unavailability of sex for men is then used by the women to control the men and as a means to gain power over the men. Naturally the men feel it necessary to develop other forms of power to counter this power and as a source for bartering.

The Other Side of the Coin, p. 113

Men explore women's sexual signals.

Although almost any woman at some time in her life will likely have a man explore her possible interest in sex with him, the woman who advertises, however subtly or unconsciously, will be having far more men probing to test the validity of the signals and reacting in angry ways when the "yes" signals are denied by the woman in self-righteous indignation.

Just as a person in a desert will check out signs that it might rain, or that a source of water might be found, men do test out women's signals hinting at sexual availability.

The Other Side of the Coin, pp. 146–7

Sex to men is like jobs to women.

Some time ago, on the same day, a fellow said to me: "I don't see why women think jobs are so important," and a woman said to me: "I don't see why men think sex is so important." Here we see serious insensitivity, each one failing to recognize that relative unavailability increases importance, just as water becomes far more important in a desert.

It's time that women recognize that sex, because of this unavailability, is far more important to men than it is to women. So sexual teasing and manipulation is a far more serious provocation, more serious violence, than most women seem to recognize. The best analogy I can think of is when men dangle job promises or manipulate women with threats to their jobs—behavior which certainly enrages women.

So Why do Rapes Occur?, 4/79, The Humanist, OSOTC, p. 242

Same rules for men and women?

Of course, too many feminists want to have it both ways. Numerous times I have heard "we should have a right to wear whatever we want to wear" from "well endowed" women feminists. And since jobs are in short supply for women, and sex is in short supply for men, I agree that women should have just as much right to wear whatever they please as men should have the right to employ and promote whomever they please.

The Other Side of the Coin 149

More sex means more love.

In the December, 1973, issue of **Reader's Digest**, the idea was expressed that: "A man gives love in order to get sex, while a woman gives sex in order to get love." I believe there is substantial truth to this statement.

However, this does not mean, as we have been taught to believe, that men are more animal and lower than women. Rather, it is primarily the result of the ready availability of sex for women and the unavailability for men.

If the situation were reversed so that sex was readily available to men and unavailable to women, then I predict that women would give love in order to get sex, and men would give sex in order to get love.

The Other Side of the Coin, p. 155

Sexual desert/sexual rain forest.

I use the terms sexual desert and sexual rain forest to describe men's situation and women's situation, respectively. I think people understand that in a desert, water is very scarce. People have a tendency to really search for water and to really focus on having enough water.

If you look at male behavior, you find that men behave towards sex much the way people behave towards water in the desert. If you look at the way women behave towards sex, it's much like the way a person would behave towards water if they were living in a rain forest. These behaviors are a result of the unavailability of sex for men and the excessive availability of sex for women.

Writings, 1/91

Let's hear it for sex objects.

Viewing a person of the opposite sex as a sex object is a perfectly natural behavior. I think feminists should wake up to that reality. A baby looks on its mother primarily as an object whose purpose is to provide pleasure to the baby. This is perfectly normal. The baby, of course, ultimately outgrows this stage and develops a more mature perspective. The problem we have with sex objects, as I see it, is that men rarely move beyond that stage, and women rarely even reach it.

The Other Side of the Coin, p. 157

Women control sex.

Our culture places the man in the role of the sexual initiator—the one who seeks sex. This puts the woman in the driver's seat—**she controls sex** and thereby controls and manipulates the man or men she associates with sexually. Of course our society's "common knowledge" is that men "use" women, but this is merely a misstatement that only helps to mask the reality.

Women's Liberation—Another Viewpoint, Wisconsin Alliance Newsletter, Spring/71, OSOTC p. 212

Male animal courts female angel.

So here is the scenario. The girl is taught to deny her sexuality and to say no. The result for the girl and the young woman is that sex is always being pushed on her—sex, she is taught, she should not desire. She knows it is available almost anytime, just by not saying no, or even by not resisting enough. On the other hand, the fellow knows sex is not readily available to him. He knows he must pay for it by buying a date and by leading the girl along properly step by step, being sure to take the steps in the proper order,—knowing all along that the whole thing can blow up at any time—if he doesn't proceed by the proper steps, if he moves from one step to the next too soon; and finally right at the payoff she may balk, in spite of all the time and money he's spent on her and in spite of doing all the steps properly. The whole game seems designed to assure that men will end up being enraged. And then finally, after all of this, there's the attitude that as a man he is just so much more an animal—so gross, so vulgar, so much below her as a near angelic woman. After all she doesn't want sex, and she may give in only because he wants it so badly.

The Other Side of the Coin, p. 91

Women should meet dependency needs they create.

Women often ask me, rhetorically I'm sure, if men have some right of access to women's bodies. I'm inclined to answer that question, even if it is rhetorical, with a qualified yes.

If women were ready to give up their sexual power over men and were prepared to encourage men to develop their emotional and affectional capabilities instead of striving to keep men dependent on women, and if women were prepared to encourage men to turn to each other for their sexual and affectional needs instead of striving to keep men dependent on women, then I think the answer would be no. But as long as women promote men's dependence on women for these needs, then men certainly have some right to expect women to meet these needs.

One can reverse the genders and state that as men strive to keep women dependent on men economically, then women certainly have some right to expect men to meet women's economic needs.

The Other Side of the Coin, p. 170

Chapter 17: Sexual Assault and Harassment

Women's assets at risk daily.

Since sexuality and physical attributes are the most visible forms of women's assets, they have little choice but to carry these assets around with them all the time. Men, on the other hand, can take their major asset out of their pocket and put it on the dresser or even into a bank. The result is that in day-to-day functioning, women are more at risk than are men because most if not all of women's assets are at risk, whereas men usually carry around only a small part of their assets. Of course women could change this by redefining what their assets are.

Achievement or Existence, 10/86, Legal Beagle, WBH

Same game, different rules.

Sexual harassment statutes are currently seriously biased against men. These statutes treat as acceptable women's sexual power over men and treat as violations only men's responses to that oppression by women. Sexual harassment often involves negotiations between equals. This occurs when a man has gained enough economic power to match the woman's sexual power. At that point the man may strive to initiate an exchange of benefits—i.e. I'll give you economic benefits in return for sexual benefits for myself. Our laws punish the man for abusing his economic power but do not punish the woman for abusing her sexual power. For example, if the woman begins to negotiate with a man to trade what she has for what he has (sex for economic advancement), that may not be seen as socially acceptable, but it is not treated as illegal sexual harassment.

Sexual Harassment is a Two-way Street, 1/87,& Statement to Wisconsin Equal Justice Task Force, 12/89

Superior woman has second thoughts.

Alternatively, the woman decides after she gets home from the date that, morally superior as she is, she could not have agreed to be sexual; so to maintain her morally superior attitude, she concludes she must have been raped. It is no accident that society in the past has tended to treat rape charges differently than other accusations. The reality is that a far higher percentage of such accusations are questionable or outright false. In fact a recent study done by the military found close to half of the investigated accusations of rape were unfounded.

Guest Editorial, Capital Times (Madison, WI), 12/10/87

Sexual assault due to powerlessness.

Feminist writers insist that sexual assault is not sexually motivated but is rather an expression of power and domination. I propose on the contrary—that it is an expression of powerlessness. Men are dominated and controlled sexually by women. So sexual assault is a striking out at and a rebellion against that domination and control in about the only way most men can do so—physically.

The Other Side of the Coin, p. 178

Men's passive-aggressive behavior.

Women get from men pretty much what they want from men—except when rage at women's oppression of men explodes in physical violence or is expressed in other ways, like passive-aggression. Passive-aggression is, after all, an expression of powerlessness. For men it is powerlessness in the face of women's perceived "spiritual" superiority.

The Other Side of the Coin, p. 182

Fair punishment for sexual assault.

Our current laws punish sexual assault almost as severely as murder. A more fair level of punishment might be the equivalent of theft of services, or of aggravated battery or assault, since the woman has indeed been a contributor to the offense. Current intense punishments are based on acceptance of women's moral superiority and a resulting sense of horror that the woman has been lowered to moral equality with men. It is also based on the attitude that women's "market value" has been lowered because she is "tarnished merchandise". This is only one of the many ways that women perceive and use sex as a commodity.

Statement to Wisconsin Equal Justice Task Force, 12/89

How women can stop rape.

Rape, as Nicholas Groth found (**Men Who Rape**, 1979, Plenum), is usually done by men who have been emotionally abused by women's self-righteous viciousness to the point that they have had their self-esteem destroyed. If people were not so intent on blaming men, they might realize that rape is a response to women's attitude of self-righteousness. So rape can best be prevented by having women come down off their pedestal of self-righteous moral superiority. As a part of this, women can begin initiating sex equally with men, instead of leaving the burden entirely on men and then blaming men when they don't do it right.

Guest Editorial, Capital Times, (Madison, WI) 12/10/87, WBH

Chapter 18: Sexual Barbarism

Sexual assault of infant boys.

From the moment a blue blanket is put on a boy, he is touched less, he is permitted to cry more. And within the first week of his life he has been sexually assaulted, with his mother's acquiescence. I refer here to that medically unjustifiable sexual assault called circumcision performed primarily because "everybody does it!" and because the doctor wants to pick up another quick and easy hundred dollars or so. **Barbaric!** I would venture to guess that, if given the choice, 99.9% of non-Jewish men would refuse to be circumcised. But few men are given that choice. There is a claim that circumcision lowers cervical cancer rates because the wives of Jewish men have less cervical cancer than do other women. Since most non-Jewish men are also circumcised, the argument is nonsense. A much more realistic argument could be made for the eating of Kosher meat or the avoidance of eating pork.

The Other Side of the Coin, p. 90

Males should have right to choose.

Most boys have been sexually assaulted by the time they are one week old. This is the violence of circumcision. Although people continue to argue about the subject, most support for circumcision seems to be aimed at justifying it because it is the current standard practice, and it provides some quick easy money for the doctor. There appears to be no substantial basis for any health claims. However, even if there were some such basis, it still seems reasonable to let the boy have the right to decide whether he will be sexually violated or not. That decision certainly should not be made any earlier than the age at which females are considered to be adult enough to choose to have sex with an adult male. In Wisconsin that age is set at 18 years.

Writings, 8/91

Chapter 19: Sex as Commodity

Feminists protect market value of sex.

Girls are taught from birth that sex is their most valuable asset or commodity and that scarcity increases its market value. Thus when a woman wants to say she is not sexually free, she will say, "I'm not cheap", i.e. she sells her sexuality for a high price. The lowering of commodity value is why feminist leaders are so outraged by sexual assault against women. If they were outraged about dehumanization, these feminist leaders would of course be equally outraged at sexual assault against men, including men in prison. But this suffering by men is pretty much ignored by them.

Writings, 8/91

Why women object to prostitution.

Renowned feminist writer Carol Cassell affirms that even today women perceive sex as their most valuable asset. It is because of this that prostitution is opposed so vehemently by women since it reduces the market value of sex for other women. As a result, this is one of the few areas where women do not receive more favored treatment by the legal system.

Report on Wisconsin's Equal Justice Task Force, 1991

Getting the best deal.

The attitude that women don't need men, or sex, does provide very important benefits for women. It is a major source of women's power. A younger woman usually has so many men "after her ass" that she can pretty well pick and choose among male economic objects for what appears to be the best deal and then will expect royal service with no effort on her part. I don't see that as any kind of a good deal for the man, unless being seen in public with a very attractive woman is worth the price. To me it isn't.

The Other Side of the Coin, p. 77

Experimentation confirms market value.

There is a lot of furor lately about teen-age girls taking the initiative in relationships and even sexually. Instead of encouraging and praising them, their elders generally put them down and make intense efforts to indoctrinate them to "proper feminine behavior". Actually, quite a few young women also do a bit of experimental initiative taking. But most women soon learn the economic value of their sexuality and how scarcity enhances its market value, and they then choose to avoid any future pain associated with the risk of rejection.

Writings, 9/91

Relationships improve when sex bartering ends.

Some women ask: "If we don't have sex as a commodity to barter with, how can we attract men?" In response I suggest that it may be surprisingly easier for men and women to relate together harmoniously if sex is not used as a bargaining item whose value is enhanced by scarcity. For one thing, men would not find it necessary to develop economic bartering assets. Personality, personal interests, and compatibility would then become more important qualities to consider in choosing one's long-time companion or even close friends.

Writings, 8/91

Women assuming responsibility.

Feminists recently censored a song by a female musician in which the female singer says her "**no**'s" might actually mean "**yes**". This song depicts a regular experience by men, but it is not politically correct to acknowledge this. The ambiguity women experience is a result of the conflict between their desire to enjoy sex and their lifelong training to see sex as their most valuable commodity which is enhanced by scarcity, and their moral superiority. Actually the most serious political error women commit is in creating conditions that deny men the chance to say no. Women seriously committed to equality and to relating to men in a caring way should assume responsibility for.initiating dates and sex at least half the time rather than leaving all the responsibility on men, and then blaming men if they don't do it right.

Writings, 8/91

At last! We can stop Date Rape!

Dr. Schenk's Guaranteed Formula for Eliminating Date Rape:

WOMEN: Take the initiative in arranging and paying for dates.

WOMEN: Take the initiative in initiating sex, especially the first time and ideally on the first date, while respecting men's right to say **NO**!

WOMEN: Treat men as human beings rather than primarily as billfolds or sources of present and future money to be spent on women.

WOMEN: Above all, give men the respect

and dignity and the caring and sensitivity that you expect from men.

MEN: Now that women no longer treat you as less than human, treat women with the respect and dignity and the caring and sensitivity you expect to receive from them. This includes using a high quality condom when sharing sex.

MEN: Welcome and encourage women's taking the initiative.

MEN: Don't tease or lead women on sexually and then refuse to be sexual with them.

That's it! It's that simple! And it's guaranteed! Now some people might say that this involves putting most of the burden on women. But then most of the burden has been on men in the past, so this is just a bit of affirmative action. Surely we all support affirmative action! Also, it is still less stressful for women because they do not suffer from the sexual deficit that men do. Actually, if women treated men as human beings, with respect and dignity and mutual sharing of responsibility, it is likely almost all forms of rape would be eliminated with little additional effort.

Don't want to stop date rape, just decrease it?

Now that you know how to eliminate date rape, if you've decided this is too difficult but you do want to reduce it, that can be done also. The formula for that is as follows:

WOMEN: Pay for your share of the dates, and treat men like human beings.

WOMEN: Recognize that sex is in short supply for men and so is more important for men (like water in a desert), so sexual teasing is an intense form of violence to men that women should avoid committing.

WOMEN: Make it crystal clear from the beginning if you do not want to have sex on a date, and then don't equivocate later.

WOMEN: Do not wear suggestive or provocative clothing if you don't want sex (i.e. don't advertise).

WOMEN: Do not give double messages about any sexual interest if you don't want sex.

WOMEN: Do not engage in any kind of kissing or making out if you don't want sex.

MEN: Be honest and up front about any sexual interest you may have. If you have a sexual interest and the woman doesn't, don't go out with her.

MEN: Treat women with the respect and dignity you expect to receive from them.

Dr. Schenk's Guaranteed Formula for Eliminating Date Rape, 8/91

Chapter 21: Gender Reversal

Different gender, same experience.

I find it interesting how often I can take writings of feminists and other oppressed groups and just change the group to "men" or reverse the sexes and find that their statements are still true. Below are some examples.

"It must be repeated once more that in human society nothing is natural and that (man), like much else, is a product elaborated by civilization. The intervention of others in (his) destiny is fundamental: if this action took a different direction, it would produce a quite different result. (Man) is determined not by (his) hormones or by mysterious instincts, but by the manner in which (his) body and (his) relation to the world are modified through the action of others than (himself). The abyss that separates the adolescent girl and boy has been deliberately opened between them since earliest childhood; later on, (man) could not be other than what (he) was made, and that past was bound to shadow (him) for life. If we appreciate its influence, we see clearly that (his) destiny is not predetermined for all eternity." (de Beauvoir, S. The Second Sex. 1952.)

"Any (woman) who profits from the exploitation of (men), profits in any way at all, is responsible for the exploitation of (men)...you cannot judge equally things that have different contexts, and that no (woman) knows what it is like to live as a (man)." (French, M. The Bleeding Heart. 1980.)

"It is a mistake to see (women) as pitiable victims or vessels to be 'saved' through (male) self-sacrifice." (Daly, M. Gyn/ecology. 1978.)

"By rejecting the false self for so long imposed upon us and in which we have participated unwittingly, we (men) can forge the self-respect necessary in order to discover our own true values. Only when we refuse to be made use of by those who despise and ridicule us, can we throw off our heavy burden of resentment. We must take our lives in our own hands. This is what liberation means. Out of a common oppression (men) can break the stereotypes of feminine-masculine and enter once more into the freedom of the human continuum." (Roszak, B. in Masculine/Feminine, B. & T. Roszak, 1969.)

Perhaps for a long time (women) will need a kind of compensatory education in the things about which their education as (females) has left them illiterate.... It means that we begin to expect of (women), as we do of (men), that they can behave like our equals without being applauded for it and singled out as 'exceptional.'" (Rich, A. Of Woman Born. 1976.)

'What is it like to live as a (male) in our

society, at a time when (masculinity) and (men) are increasingly disparaged? How does being (male) come to dominate a person's social identity and self-concept?" (Millman, M. Such a Pretty Face, 1980.)

The Other Side of the Coin, pp. 167–8

Chapter 22: Men's Feelings and Needs

Men's first feelings may be painful.

We need to look at the suppression of men's feelings and how we can change the conditioning that creates this. My experience indicates that we will have as great a difficulty in getting recognition that our feelings are important as women have in getting their ideas recognized as important. Also, one of the problems I experienced in becoming aware of my feelings was that the first intense feelings I became conscious of were not joy and acceptance, but rather they involved hurt and pain and such things as loneliness, anger, rejection, and sadness. Likely other men will have a similar experience.

The Other Side of the Coin, pp. 135–6

Violence done to male children.

There are some positive things to be said about the way boys are raised in our society, such as the permission to explore, to be physically active, and some of the other aspects associated with men's expected need to achieve; but the attitude of spiritual inferiority ("He's just a boy so you can't expect anything better!"), the structure of the school environment which values compliance and results in boys experiencing most of the reprimands and punishments (enhancing the perceptions of inferiority), the tendency to expect toughness and to give less touching, indeed the whole training to be out of touch with our feelings and to accept the devaluation of seeing ourselves as legitimately being potential cannon fodder, are examples of violence which have devastating effects on male children and on the men they become.

The Other Side of the Coin, p. 90

Men must deal with rage.

Any men's consciousness raising group that doesn't deal with men's rage at women is dealing with a veneer and, if anything, is adding to men's already excessive feelings of shame.

Time for a New Phase in the Liberation Struggle, 1975, OSOTC, p. 225

Violence with intimacy.

The major reason that men fear intimacy is because of the enormous amount of (emotional and sexual) violence that they experience when they get into an intimate relationship with a woman. And it is perfectly reasonable that men should respond that way—just as it is perfectly reasonable for women to respond negatively when they are living with a man who is violent to them.

Overcoming Men's Fears of Intimacy, workshop at the 16th Midwest Regional Conference of the Association of Humanistic Psychology, 1986

Male attributes.

I believe when we are finally really able to come to grips with the whole picture, we will find that the present "masculine" attributes, like courage, assertiveness, independence, and strength, will contribute a share of the virtues of the new-human-person-which-we-hope-to-become which will be equal (but obviously different) to those of the female.

Time for a New Phase in the Liberation Struggle, 1975, OSOTC, p. 223

Homophobia due to shame.

Intimacy with other men is seriously impeded by men's feelings of competitiveness. But the competitiveness, overachievement, and other characteristics of macho behavior are all responses to feelings of shame. And men, of course, being competitive with each other, find it difficult to be intimate with each other.

Overcoming Men's Fears of Intimacy, workshop at the 16th Midwest Regional Conference of the Association of Humanistic Psychology, 1986

Men-only groups.

Since women control society's mores and morals, men naturally strive to get away from women into male-only groups where they can escape the shaming influence of women. Women are succeeding in gaining government enforced access to more and more of these groups, thus destroying their purpose for existence. We can only hope that the all-male groups men develop to replace the infiltrated groups will most frequently be male-affirming, socially beneficial groups.

Writings, 9/91

Intimacy puts men one down.

Because men have this feeling of defectiveness, of not being loveable, when a woman loves a man, there is the feeling that she is doing him a favor. And as a result a man is in a one-down position to a woman when intimacy between a man and a woman develops.

Overcoming Men's Fears of Intimacy, workshop at the 16th Midwest Regional Conference of the Association of Humanistic Psychologists, 1986

What do young men need?

Above all else, young men need to be affirmed, to hear and to be able to accept that they are worthy, that they deserve to be treated like human beings. One way that this can be achieved is by having older men serve as mentors for young men. I believe Robert Bly is correct when he says that we older men need to bless and celebrate younger men.

Writings, 2/91

Chapter 23: The Men's Movement

The men's movement different from the women's movement.

For a long time I've realized that the men's movement would be different from the women's movement. Women, after all, are struggling with overcoming the message that they are inferior in the "doing" or achievement areas. This requires helping women overcome their fears and feelings of incompetence.

Men, on the other hand, are struggling with overcoming the message that we are inferior in relationships, in morality, in "being". In other words our negative message says we are defective human beings because we are male. Overcoming that lie requires us to energize and honor the feelings and emotions we were forced to suppress in order to survive our childhood. It requires us to learn to recognize the lovableness within us, however hard we have suppressed it. That is the first important task of the men's movement.

Writings, 10/91

Supporting men.

Unfortunately, or perhaps I should say fortunately, most men have enough shame put on them already so they are not prepared to buy into a men's movement which is based on further shaming of men. Most men, I believe, are waiting for a men's movement which concentrates on addressing men's issues and men's oppression while not putting down the equally important women's movement.

The Other Side of the Coin, p. 60

Support men who rape.

One of the most important things we men need to do is to learn to support each other at times of adversity. This was brought home to me very powerfully recently when an associate was accused, tried, and convicted of a rape I am convinced he did not commit. Financial appeals to men active in the men's movement produced essentially no help. And most local men not only offered no support whatsoever but also in fact shied away from any identification with the case.

This contrasted dramatically to the broad community support a local woman got a year or so ago after she gunned down her ex-husband. As I've noted before, women's violence to men is treated as self-defense and so as justified, or it is simply ignored.

It's time we men come to recognize that men's violence to women is just as much a reaction to women's violence to them as is women's violence to men. So that even if a man does commit a rape, we should support him even though we do not approve of his violent actions. We would, after all, support a son or a brother who broke the law, even when we do not approve of his behavior.

The Other Side of the Coin, p. 137

Supporting our brothers.

Here is the reason there is no significant men's movement in the U.S., or anywhere for that matter! As men, we are all of us so burdened with **Shame of Maleness** that we pull away from any man who is **even accused** of doing violence to a woman. We treat him like a pariah. We refuse to support him either emotionally or financially. How can we expect to develop the kind of supportive network we need for an effective movement when we reject and abandon our brothers in this way? Let's realize that these expressions of violence do not occur in a vacuum. When men commit violence, they are usually responding to violence done to them.

It is simply essential that we as men learn to support each other, emotionally, financially, and all other ways, regardless of the extent of our warts, or even our open, festering sores. To do otherwise to is judge our brothers as less than human, which is to buy into the perpetual myth of women's moral superiority, which damages us all.

Letter to Transitions, 5/87, WBH

Men's issues harder to perceive.

Men's gender related limitations and inequities are obviously different than the ones women experience. More of a problem is their greater subtlety, their lack of concreteness. It's easy to see and get angry about lower pay and unavailability of advancement opportunities which women experience. But it is much harder to perceive and address the conditions that set us men up to accept being cannon fodder and subjecting ourselves to excessive, life-shortening stresses.

The Other Side of the Coin, p. 5

Men's Studies Programs needed.

One very important way to assist the elimination of anti-male biases will be the development of Men's Studies Programs which can more thoroughly study and document these matters, just as Women's Studies do this for women already. Feminists will undoubtedly scream that this is an attack upon women, that it denigrates women. (This is true if you accept the belief that women are superior to men, since it proposes viewing women as equal to men.)

Statement to Wisconsin Equal Justice Task Force, 12/89

No one deserves abuse.

It is only when men have experienced intensely outrageous abuse such as often occurs in divorce that most men can fight back. And even then, after the abuse has been reduced or even found to be irreversible, these men generally fall back into the same old pattern of shamed acceptance of being treated as inferior. **That is the reason so few men continue in men's movement activities after they deal with their acute abuse.** They accept the day-to-day abuse as deserved because they are inferior males, and if deserved, then simply accepted.

How About the 20th Century?, The Liberator, 1/90

Don't criticize your superiors.

Men are generally intensely fearful of saying anything critical of women or of even being associated with a man who says anything critical. It reminds me of how blacks behaved in Georgia when I lived there. Blacks didn't dare criticize superior whites, and they dissociated themselves from any one who did. Even most men involved in the men's movement respond this way. Thus **Wingspan,** billed as the journal of the male spirit and the largest men's movement newspaper, refuses to print anything in any way critical of women's behavior, even though the wounding of men's spirit results mainly from oppression by women. Even Robert Bly has responded very negatively to me when I have made comments critical of women's behavior. It seems hard to believe that we can have an effective men's movement as long as men are so terrorized by the thought of offending "superior" women.

Writings, 6/91

What men can do.

A man can learn to be conscious of the continual shaming he experiences from women and can strive to challenge and confront that shaming. But the most important thing we need to do is to learn to love ourselves and to realize that we are lovable—as indeed we are. From birth we have been told the lie that we are defective because we are male and therefore we are unlovable. Having heard this lie so long, we finally come to believe it. We internalize it, we accept it as true. Now our job is to unlearn that lie. That isn't easy because we learned it so early and we've had it repeated so regularly that it just seems true.

Affirmations such as: "I love myself" and "I'm loveable" can help. However, for most men it is likely that involvement in a group of men committed to affirming men is necessary to help develop the self-love and self-caring necessary to move beyond the self-doubts and sometimes self-loathing and so to become fully human. Groups like the **New Warriors** (phone: 414-964-6656) are particularly helpful for surmounting this shame. Once a man has developed some self-love, he needs to commit himself to helping other men. This involves inviting other men to join in the healing. It also needs to involve a commitment to working on reversing the numerous anti-male laws that have been passed at all levels of government over the past 20 years.

Writings, 8/91

Chapter 24: Equal Pay for Equal Work

Are women really underpaid?

I certainly agree that women should receive equal pay for equal work. However, it is not at all clear to me that the average woman should be paid as much as the average man is even if they work at the same job, because women appear to be generally less productive than men are. Women spend more time on the telephone. They spend more time in the restroom. This is why feminists are demanding relatively more restroom facilities for women in public places. Also women expect men to do the dirty, dangerous, and heavy parts of the job. In addition, full time male workers average 3.5 hours more time on the job per week according to the Department of Labor. And men's shame drives them to over-achievement and success. The average woman now earns 75% as much as the average man and she may well be overpaid if we use productivity as a basis for pay. If we justify paying women more merely because they are female, as feminists seem to desire, that is sexual discrimination.

Writings, 7/91

Women fear better paying jobs.

It is legitimate to ask if the average woman should be paid as much as the average man. For one thing, women's fear often prevents them from taking the risk of seeking better paying jobs and/or job advancements. Thus, for example, most recent data show starting pay for women chemists is equal to that for men; but over the years their average pay does not increase as fast, until their overall pay totals only 88% of that for the average male chemist. This undoubtedly results from women's fear and lack of assertiveness which tends to leave women satisfied with less stressful and demanding jobs.

Writings, 8/91

Sexy clothing no substitute for competence.

The continued flourishing of the fashion and cosmetic industries shows that women even today are depending on their bodies and their looks as their primary assets. One does not need clothing that calls attention to and enhances one's sexuality to achieve at a job. Women who wear clothing on the job that calls attention to their sexuality are obviously hoping their attractiveness will compensate for their lack of competence.

Writings, 8/91

Chapter 25: Marriage, Divorce, and Custody

Women need to oppose anti-male legislation.

Any woman who would like to share her life with a man should be very, very concerned about the immense amount of anti-male legislation that has been passed in this country since feminists gained control over the past 10–20 years. Today a man risks his life when he even dates a woman and is alone with her, as men imprisoned on false accusations of date rape can attest. Marriage has become a contract where men gain obligations and women gain rights and privileges that continue even if the couple divorce. Not surprisingly, the number of men aged 30–34 who have never married has tripled in the past 20 years. A high percentage of these men will never marry if past experience holds true. As a result a lot more women won't marry because they can't find a man who is willing, and men who become eligible through divorce are quickly remarried.

Writings, 6/91

Divorce teaches men.

Divorce generally is one of the most intense emotional traumas a man will experience in his life. The typical fellow is ill-prepared for the stress because society has trained him to deny his feelings. And it has trained him to the supposedly superior status of being a money-providing machine and little else.

Divorce: His & Hers, Milwaukee Journal Mag., 4/8/84

Marriage has become an onerous contract for men.

Every man and woman sign a contract when they marry. Either they write their own contract or they accept the contract written into the law books. And the courts may later nullify the couple's contract anyway. Nonetheless, any man who unquestioningly accepts the state's contract, maybe any woman also, is acting in an incredibly foolish way.

In the traditional marriage contract, women give men pussy in return for the man's supporting them. When wives began going to work, men feared that they were failing in their part of the marital contract, in which case women might refuse to honor their part of the contract also. That has become increasingly true.

The traditional benefits men gained from marriage have been largely eliminated. However, the responsibilities and obligations remain and become more onerous beyond divorce. Thus the divorced man is expected to continue providing financial support to the woman who has robbed him of his children and who no longer provides him with pussy or any of the other traditional benefits of marriage. It is really too bad that commitment by men has become such an onerous burden. Men, as much as women, need to care for and be cared for by at least one other person. And feminist-instigated laws are making men's responses to that need of men increasingly risky.

Writings, 9/91

Men choosing not to marry.

In the past men have devoted their lives to supporting their special woman, basing much of their self-worth on how well they could provide her with ease and luxuries. Recently, more and more men have rejected this raison d'etre. As a result, women can no longer depend on having a man to play the knight in shining white armor, supporting and protecting them. Consequently, feminists have turned to the government to play this role for them by forcefully extracting from men the support men are now less willing to give voluntarily because the trade-offs men used to obtain in return have been eliminated at the insistence of feminists. They have also shamed male legislators at all levels of government into passing anti-male legislation of increasing ferocity, including divorce and custody and sexual assault and harassment laws. This is guaranteed to further increase the numbers of men who choose not to marry. I wonder if that is what most women really want.

Writings, 3/91

Chapter 26: Further Observations

Women love success objects.

Typically a woman doesn't fall in love with a man who has a lower economic potential than she has. It seems then that the feature women find most lovable in men is their money earning potential. In other words a woman first evaluates the man's status as an economic object, and if this measures up to her evaluation of her beauty/sex object potential she may permit herself to "fall in love". Even so, taught from birth that he is defective and therefore unlovable because he is a male, the man will likely be overwhelmed by the fact that a woman could find him lovable and so he may well seek a permanent relationship with this "unique" woman.

Writings, 8/91

Women in combat?

Should women in the armed services be **permitted** to participate in combat? I think the answer has to be a resounding **no!** That just gives women one more special privilege. Women in the military should be **required** to participate in combat as much as men are required to do so.

Unpublished letter to editor, 5/91

Vietnam war statue for women.

Vietnam combat nurses are fighting to have a statue of a female erected at the Vietnam War Memorial in Washington. And they are certainly right that women should get recognition commensurate to their sacrifices in that war. Now let's see, there were 58,000 men and eight women killed. Since there are three life-sized statues of men, that's 216 inches of statue for 58,000 men. If my calculations are correct, and if the eight women are to get a fair and equal representation, they should have a statue almost 0.03 inches high. Surely we can find space somewhere around the memorial for a statue of these dimensions.

Letter, Capital Times, 8/86, WBH

The new "niggers".

Why is it that the high proportion of blacks executed in the U.S. causes shocked reactions but that the far higher proportion of males executed (now 142 men to one woman) does not? Is it that being male is perceived as being even more evil than being black? Or aren't we supposed to ask this kind of question?

Letter, Capital Times, 12/86, WBH

Stop dehumanizing men.

By calling a man a "rapist", a "perpetrator", a "murderer", or whatever as a result of his doing violence to women, one defines that man on the basis of one negative action or series of actions. This dehumanizes the man, which serves to make him unworthy of empathy or other humane treatment. We can then focus empathy solely on the "victim" and do not need to investigate any contributory behavior by her, or by women as a class for that matter. For example, we can ignore the perpetual emotional abuse that women heap on men through dangling sex, through seeing men primarily as economic objects, through treating men as morally/culturally inferior and "looking down their noses" at men, by expecting men to protect them, by the verbal abuse that so many wives use to emotionally devastate their husbands, and by the widespread physical violence by women which is generally ignored or condoned.

Feminists' claims that women do not contribute in any way to the violence men do to women is simply a lie, meant to maintain the mirage of women as innocent victims. As Kahlil Gibran says in **The Prophet**, "The righteous is not innocent of the deeds of the wicked, the guilty is oftentimes the victim of the injured and the condemned is the burden bearer for the guiltless and unblamed." The sooner we can realize that this applies to women in relation to men equally as well as to other situations, the sooner we can begin to reduce the interrelated violence and move towards improved relations between men and women.

Writings, 9/91

"Women's equality" another name for privilege.

We hear a lot of complaints that men do not do their share of housework. Yet, when it comes to hard physical work, women are conspicuously absent. Heavy physical work can be exhausting. In fact, after 10–15 minutes of intense physical work I can be exhausted for a couple of hours. Does my recovery period count toward total work time? Perhaps there are still a few super-macho males who would try to prevent women from doing heavy work; but if women were serious about equality, they would demand the right to do their share of the heavy physical work. But, then, it appears that most women want equality only up to the point where it becomes inconvenient.

Letter to Ann Landers (unpublished), 2/90

Men do, women are.

Women exercise a fundamental control over men due to women's perceived moral superiority and through control of socialization. Women's control of sex through maintaining a sexual scarcity for men is another source of power and control for women. Women use this control to be able to get their self-worth based on existence rather than on achievement, which is left for men. (In my opinion, neither is acceptable as we should have self-worth based simply on being alive.) The cosmetic and fashion industries are based on aiding women to promote themselves as worthy. Men, however, must achieve to get the same acceptance and approval.

How About the 20th Century?, The Liberator, 1/90

When privilege becomes obligation.

In the past, household work was so hard and time consuming that a woman was an equal partner with her husband just by doing the domestic tasks. Then men developed an array of labor saving devices for the home, and women's work became so easy that they no longer felt like equal partners by doing just the domestic tasks. So women demanded the **privilege** to work outside the home. Shortly after that time, many third world countries rebelled against the stealing of their wealth by American multi-national corporations and organized to raise the price of oil. This started a decline in real wages for Americans which has increasingly forced women into the market place in order to maintain living standards. As a result, women's privilege to work outside the home has frequently become an obligation—and many women don't like that.

Writings, 9/91

Beware the matriarchy.

Feminists regularly condemn patriarchy, which they see as a system for dominating women. Actually they are condemning something better called hierarchy, since both men and women can move into the dominant slots of the system, though overall the power positions do contain mostly men. What feminists choose to ignore is that when patriarchy develops, a parallel system of matriarchy also develops. In the most intensely patriarchal systems one will also find the most powerful matriarchal systems. It is the system that controls the covert, the family, social morality, behavior and relationships. This system is totally dominated by women. Any time patriarchy is condemned, this parallel dominance system of matriarchy also needs to be equally condemned. Yet feminists choose to ignore women's power system and condemn only men's compensatory power system. This is just another evidence that feminists are interested in power and their claims to be seeking equality are just a facade.

Writings, 8/91

What do women want?

Many years ago, Freud asked the question: "What do women want?" More than one feminist has told me the answer is: "Everything." They want men to give up their areas of power to women while women retain their current sources of power. For these feminists, equality means having men be totally enslaved to women rather than the only partial enslavement men currently experience as a result of women's control of relationships and morality.

Writings, 9/91

Eliminate fucking and screwing as pejoratives.

The use of the words "fuck" and "screw" as pejoratives needs to be examined. The negative connotations of these words result from the perception that women are above "that sort of thing", i.e. sex. If one is so superior that one doesn't lower oneself to fucking or screwing, then to be fucked or screwed is naturally a very bad thing. So it is quite reasonable to extend the words to mean any bad experience—"I got fucked!" "Screw you!" etc. Thus use of these words in this way is dependent on a perception of female moral superiority. So persons seriously committed to male-female equality should seriously consider eliminating this usage from their vocabulary.

The Other Side of the Coin, p. 42

Putting sex in its place.

If sex can be abandoned as a source of power for women, men's sexual needs can be more effectively met, in which case men's sexual urgency will dissipate. Then both men and women will be able to enjoy, even savor, the touching, the affection, the caressing which make of sex a still essential and important but far less urgent aspect of male-female relations.

The Other Side of the Coin, p. 85

Feminists not smart in allying with male feminists.

I believe feminists do themselves a serious disservice by accepting "male feminists" as their chief male allies. They would do far better to work with men who are able to help women see how women contribute by their sex roles to our mutual problems through acting as society's agents in the oppression of us men. It is likely true that these men will often come across as abrasive and enraged. So it will be less comfortable in the short run, but it also will produce much more desirable and extensive social changes.

The Other Side of the Coin, p. 66

Chapter 27: Where Do We Go From Here?

The problem.

In the areas of achievement or "doing", men have been seen as the standard, and women have been considered to be unable to measure up simply because they are female. This has caused feelings of incompetence in women, which add to the feelings of fear which women learn as children.

In the areas of relationships and morality or "being", women are seen as the standard, and men are considered to be unable to measure up because we are male. This has caused men to feel we are defective and therefore unlovable. This creates intense feelings of shame in men for being male. Boys show this shame already before they are three years old. Since it is learned so early, men are not generally consciously aware of these feelings. The shame just seems to be a part of men's life. It is this **Shame of Maleness** that permits women to manipulate and control men to get whatever women want from men. Like other women, many feminists seem to want to continue believing that this manipulative control of men by women is inherent rather than a result of men's shame feelings. Actually this shame and manipulative control are the greatest impediments to equality, to reduced gender violence, and to a peaceful world.

A Revolutionary Proposal for Ending War, 1991

The solution.

Therefore we have got to change the ways boys and girls are socialized—girls to fear and boys to shame. In the meantime we need to work to reduce the impact of these effects on ourselves. This means women need to take responsibility for developing confidence and assertiveness instead of putting their efforts into blaming and manipulating men, including manipulating legislators to pass more laws to further abuse men.

For men this means developing a sense of self-love and self-worth to replace our learned sense of defectiveness and shame. The men's movement, motored by Kauth and Bly, by Moore and myself and others has to have this as its clear and primary initial goal. This is the first essential step. That will not be enough, though, as we must also change our society's whole anti-male legal and social system which views men as "The Bad Guys" and often punishes men over 20 times more severely than it does women and which considers women's special privileges as inherent. In short we men need to work for true equality between men and women as the ultimate goal. This is the only way for the human race to survive.

A Revolutionary Proposal for Ending War, 1991

How do we achieve change?

Are women likely to willingly give up the manipulative power over men which women gain by developing the **Shame of Maleness** in us men? Since most of us tend to consider short term, rather than long term, self-interest and so tend to take the easy way when faced with choices, it is likely that most women will continue to choose shaming and manipulation until we can convince them it is in their short term best interest to change.

The way we men can best do this, and so improve relations between men and women, is by refusing to accept women's shaming messages and by refusing to associate with women who insist on continuing the emotional abuse of shaming. The substantial decline in numbers of men willing to marry indicates that this is already happening on a modest, subconscious level. Change which will benefit both men and women requires that men consciously and extensively reject the ongoing shaming and the women who do it. It is important to let these women know they are being rejected because of their emotional abuse, so there can be hope that they will change their attitudes and behaviors. Perhaps the best way to convey this message will be by consciously seeking out and celebrating women who do valiantly strive to stop their emotional abuse of us men and who strive to treat us with respect and dignity as equal human beings.

Writings, 9/91

Moving beyond intersexual violence to love.

Consciously absent from this book is any discussion of love. That is not because I think love is unimportant. Rather I believe that correctly understood, love is the most important and potentially the most powerful force on earth. But I believe our present gender related violence, induced by inferiority/ superiority attitudes, warps and all but destroys real loving in our world today.

A wise old woman defined love for me as "tender caring". That is the best definition I have found so far. Tender caring is something we can extend to anyone. It is not restricted to someone of the opposite sex. Still, for most of us, love between ourselves and a member of the other sex can become the most intense and intimate experience of love. Unfortunately the violence done by women to men and by men to women is so intense that this desirable experience is rarely attained more than just briefly, as can be attested by the high percentage of marriages that end in divorce.

What we call love too often comes close to being a mere narcissistic using of the other person for our own benefit rather than an ongoing mutual experience of tender caring.

When we have been able to largely remove from our lives the intersexual violence we learned as children and learn to treat each other with respect and dignity and tender caring, that love will transcend anything that most of us have even glimpsed in the past.

Writings, 9/91

Utopia pale by comparison.

If we can stop training women to fear and to "doing" inadequacy and can abandon the message of women as the "being" standards and so men as flawed and defective, we will create such fundamental changes in the behavior and attitudes of men and women that we will experience blissful relationships between men and women of an intensity far greater that anything we currently can even conceive of. As we approach this goal, I predict that war will also become intolerable, and the whole interpersonal dynamic will change dramatically in ways we cannot now imagine. Some people may accuse me of wild overstatement in making these claims. On the contrary, I believe the future will prove the statements suffer from understatement.

Introduction, by Roy U. Schenk, Ph.D., to: ***Healing the Shame of Maleness****, Schenk, R. U. and J. Everingham, Editors. In preparation; expected publication date, 1992*